ADVANCE AND RETREAT

PERSONAL EXPERIENCES

IN THE

UNITED STATES AND CONFEDERATE STATES

ARMIES.

BY

J. B. HOOD,

Lieutenant-General in the Confederate Army.

PUBLISHED FOR THE
HOOD ORPHAN MEMORIAL FUND.
G. T. BEAUREGARD,
NEW ORLEANS, LA

1880.

PRESS OF
BURK & M'FETRIDGE,
304 CHESTNUT ST.
PHILADELPHIA.

CONTENTS.

APPENDIX.

ADVANCE AND RETREAT.

CHAPTER I.

UNITED STATES ARMY—CALIFORNIA AND TEXAS CONFEDERATE STATES ARMY—VIRGINIA, YORKTOWN, ELTHAM'S LANDING, SEVEN PINES OR FAIR OAKS.

I RECEIVED at the age of seventeen an appointment as Cadet at West Point through my maternal uncle, Judge French, who was then in Congress. I fancied a military life, although it was not my father's choice. He occupied a high position in the medical world, and preferred I should adopt his profession; he offered me every inducement—even the privilege of completing my studies in Europe. I, nevertheless, adhered to my decision. Doubtless I had inherited this predilection from my grandfathers, who were soldiers under Washington. They were of English origin; had settled at an early period in Virginia, and after taking an active part in the War of Independence, emigrated to Kentucky, "the dark and bloody ground," where they lived in constant warfare with the Indians. One of them was married in the Fort of Boonsboro', the first fortification constructed in that State, the land of my nativity.

I entered the Military Academy in 1849, and graduated in the Class of Sheridan, McPherson and Schofield, in 1853, when I was appointed Brevet Second Lieutenant in the Fourth

Infantry. I sailed from New York in November of that year to join my regiment in California, via Panama. On my arrival at San Francisco—at that time a small city built upon sandhills and flats, and distinguished for its foggy atmosphere—I, together with one of my classmates, deemed it but proper that officers of the United States Army should go to the hotel in a carriage; but to our astonishment, on hailing a driver, we found the charge to be twenty dollars in gold. This aspect of affairs—our pay being only about sixty dollars a month—compelled us to hold consultation with our brother officers and to adopt the only alternative: to proceed on foot to whatever quarters we desired to occupy.

After having been stationed a short period at Benicia Barracks, I was directed to report for duty to Captain Judah at Fort Jones, Scott's Valley, in the northern portion of California. Colonel Buchanan was in command of my regiment, with Captain U. S. Grant as Quarter Master. It was at this post I formed a warm attachment to Lieutenant George Crook, now Brigadier General in the Army, and who has so signally distinguished himself as an Indian fighter. Although he completed his course at West Point a year before I graduated, his purse was not much longer than my own; it became therefore necessary for us to devise some plan to get along in this country of gold and extravagance. We concluded to associate ourselves with Doctor Sorrell and Lieutenant Bonnycastle in the organization of a mess, and, as we were fond of hunting and game was plentiful, to supply our own table with every variety thereof and to send the surplus to market for sale. This financial policy worked admirably, and since I had at the age of fifteen, during the absence of my father in Philadelphia, taken charge of his farm for one year with considerable success, Crook and I were led to secure land and sow a large crop of wheat. Just before the harvest, however, I was ordered in command of a detachment of Dragoons to serve as escort to Lieutenant Williamson of the Topographical Engineers, upon a surveying expedition in the direction of Salt

Lake. My duties were soon brought to a close by the receipt of an appointment as Second Lieutenant in the Second Cavalry, a new regiment organized in accord with an Act of Congress, in 1855, and commanded by Colonel Albert Sidney Johnston, with R. E. Lee as Lieutenant Colonel, George H. Thomas and W. J. Hardee as Majors. Lieutenant Philip Sheridan relieved me, and I returned to San Francisco en route to Jefferson Barracks, Missouri, the rendezvous of the regiment. At the former place I met, for the first time, in his bank, W. T. Sherman, who possessed as at present the same piercing eye and nervous impulsive temperament. Little indeed did I anticipate at that period the great theatre of life upon which I was destined so soon to be thrown as an humble actor with him and others just mentioned, and who have since become so distinguished and prominent as American soldiers.

In the early Autumn of 1855 I sailed from San Francisco for New York, via Panama, and reported for duty at Jefferson Barracks, Missouri. Soon thereafter, if my memory betray me not, I received a draft for about one thousand dollars in gold, as my share of the profit in the wheat crop cared for by Lieutenant Crook and myself. In November I marched with my regiment to Fort Belknap, Texas, which place we reached about the middle of December. Shortly afterward, Camp Cooper was established on the Clear Fork of the Brazos. Major George H. Thomas was placed in command till the arrival of Lieutenant Colonel R. E. Lee, to whom I had become very much attached at West Point where he was Superintendent whilst I was a Cadet. My relations and duties were therefore most pleasant during my service at Camp Cooper. The Government had under advisement, at this period, the construction of a fort in that vicinity; it was Colonel Lee's custom to often ride over the country in search of a suitable location, and to request each day one or more of his officers to accompany him, in order to avail himself of their views in regard to the best point at which to establish this military post. Whilst riding with him upon one of these excursions, and enjoying the scenery and balmy air as we passed over the high and undulating prairies of that beautiful region, the conversation turned upon matrimony, when he said to me with all the earnestness of a parent: "Never marry unless you can do so into a family which will enable your children to feel proud of both sides of the house." He perhaps thought I might form an attachment for some of the country lasses, and therefore imparted to me his correct and at the same time aristocratic views in regard to this very important step in life. His uniform kindness to me whilst I was a Cadet, inclined me the more willingly to receive and remember this fatherly advice; and from these early relations first sprang my affection and veneration which grew in strength to the end of his eventful career.

The latter part of that same year I was ordered to Fort Mason, situated near the Llano river, about forty miles distant from Fredericksburg. Colonel Albert Sidney Johnston was chief in command until sent to Utah. Although stationed with him but a short time, I became deeply impressed by the exalted character of this extraordinary man. Major George H. Thomas succeeded in authority; it was during my service as his Acting Adjutant that he specially won my high regard by his manliness and dignity.

After the lapse of several months, and having grown weary of the routine duties of camp life, I determined to change the scene and start on a scouting expedition in search of the red men of the forests. Preparations were accordingly made, and I left Fort Mason on the morning of the 5th of July, 1857, in command of twenty-five men of Company "G" Second Cavalry, with an Indian guide, compass in hand and supplies for thirty days. I passed out upon the plains by the head of the Llano river, and marched thence to the country bordering on the Concha rivers. After an absence of ten days and an exploration of these different streams, I discovered an Indian trail, apparently about two or three days old, and indications warranting the belief that fifteen or twenty ponies belonged to the party which was moving in the direction of Mexico, via the head waters of Devil's river. I was young and buoyant in spirit; my men were well mounted and all eager for a chase as well as a fray. It was soon apparent that we would be forced to pass over a portion of the staked plains or desert lying between the Concha rivers and Mexico; that in order to overtake the Indians we would most likely have great fatigue and privation to endure, as we could expect to find but little water during the pursuit. However, in the conviction that we could live for a short time wherever Indians could subsist, we began the chase on the morning of the 17th of July, marched about forty miles, and camped that night upon the dry plains without water or the sight of game, so frequently in view the previous day, and without even the chirp of a bird to cheer us on our journey, we knew not exactly whither. At early dawn the following morning the march was resumed; we passed during the day a water-hole utterly unfit for use, and went into bivouac that night with the same surroundings, fully fifty miles further out in the desert. Our canteens were now empty, and the outlook was somewhat dismal. At daybreak on the 19th, "to horse" was sounded and the journey continued. About noon a deer was seen bounding over the prairie, and with the sight went forth a shout of joy from the men, who then felt confident that fresh water was not very far distant. The trail had moreover become much more distinct; this encouragement, together with the hope of quenching their thirst, reinspirited the soldiers. A few hours later another pool was reached, but not of that purity which was desirable. The odor of the water was such as to oblige one to hold his breath whilst he partook of the distasteful but refreshing draught. The canteens were, notwithstanding, again filled, as well as the sleeves of all the waterproof coats we possessed. The pursuit was continued, and at dark we bivouacked after a forced march of probably sixty miles. Several of the horses began to show, by this time, great fatigue and leg-weariness. The following morning the lofty peaks of the mountains near Devil's river could be seen afar off, and all possible speed was made as we recognized that the line between the United States and Mexico was not far distant. About noon we reached another stagnant water-hole near the foot of a range of hills in proximity to the rugged and mountainous country about the head waters of Devil's river, along the banks of which stream passes the stage road from San Antonio to El Paso. Here we discovered that another party of Indians had joined that of which we were in pursuit. The deserted camp indicated that there were not less than fifty warriors in number. They had eaten one of their mules or horses, and this sign, together with others about their bivouac, bore clear evidence that the party had become formidable. The trail from this point was not only much larger, but presented a fresher appearance. The arms of the men were therefore carefully inspected, every preparation made for action, and the chase quickly resumed. The horses were much fatigued, and some of them were scarcely able to keep their places in the line of march; consequently the pursuit was not as rapid as it had been the three days previous. The march over the hills and up the mountains increased moreover their leg-weariness to such extent that about 3 p. m. I abandoned all hope of overtaking the Indians before they crossed the Rio Grande, which river was then not far distant. This condition of the horses and the thirst of the soldiers led me to the determination to quit the trail and go immediately in search of fresh water. We were at this time well up on the high and rough range of mountains bordering on Devil's river, and after leaving the trail a distance of nigh one mile, I perceived on a parallel range about two miles off a few Indians waving a large white flag apparently hoisted from a mound. Orders from Washington had been issued before I left Fort Mason, notifying all United States troops that a party of Tonkaways were expected at the reservation, near Camp Cooper, and that they would, in the event of meeting a body of our soldiers upon the frontier, raise a white flag, upon which signal they were to be allowed to pass unmolested. I therefore became convinced that these Indians were either the Tonkaways or a hostile body endeavoring by an infamous ruse to throw me off my guard, to entrap and massacre my entire party.

Notwithstanding the condition of the men and the horses, I determined to pass over upon the ridge occupied by the red men, move toward them, and ascertain the meaning of this demonstration. I had at this time but seventeen men for action, the remainder having halted in rear, owing to the inability of their horses to advance further without rest. I moved across to the opposite ridge and, as a precautionary measure, formed line and marched forward in readiness to talk or fight. Every man was armed with an Army rifle and a six-shooter; a few of us had sabres and two revolvers, whilst I was armed with a double barrel shot-gun loaded with buck shot, and two Navy six-shooters. As we passed over a mound about one hundred and fifty or two hundred yards distant from the one occupied by friend or foe—we knew not which—the flag, seemingly a sheet, was still waving aloft and a few Indians were lounging about with every appearance of a party desirous of peace.

The ground in that vicinity was rough and partially covered with a growth of Spanish bayonets which afforded a secure place of concealment. Feeling that in the event of an attack I had better chances of success mounted than dismounted, for the reason that my fighting force in the latter instance would have been lessened by the number of men required to hold and guard the horses in rear, and sharing the belief which generally prevailed in my regiment that twenty well-armed soldiers should be able to successfully engage four times their number of Indians, I continued to move forward slowly upon the immediate right of my line. When we were within about twenty or thirty paces of the mound occupied by the Indians, four or five of them advanced towards us with the flag; suddenly they threw it to the ground and fired upon us. Simultaneously from a large heap of dry grass, weeds and leaves, burst forth, in our immediate front, a blaze of fire some thirty feet in height, and, with a furious yell, the warriors instantly rose up round about us, whilst others charged down the slope in our midst, even seizing some of our horses by the bridle reins. At the same moment a mounted party attacked the left of our line with lances. Thus began a most desperate struggle. The warriors were all painted, stripped to the waist, with either horns or wreaths of feathers upon their heads; they bore shields for defence, and were armed with rifles, bows and arrows. The quick and sharp report of our rifles, the smoke and cracking noise of the fire, together with the great odds against us, the shouts of the soldiers and the yells of the Indians, betokened the deadly peril from which seemingly naught but a miracle could effect our deliverance. Each man, after discharging his rifle, drew his revolver and used it with terrible effect as the warriors, in many instances, were within a few feet of the muzzle of our arms. Stubbornly did my brave men hold their ground; again and again they drove the enemy back to the edge and in rear of the burning mass of weeds in our front, when finally the Indians charged desperately and forced our line back a few paces in the centre. Having discharged my shot-gun, I rode at once with revolver in hand to that point, rallied the soldiers, who again drove them back, whilst our horses, in some instances, were beaten over the head with shields. The contest was at such close quarters that a warrior bore off a rifle which had been used and hung by one of the men upon his saddle. Meantime the Indians as quickly as they discharged their arms, handed them to their squaws, who ran to the rear, reloaded and returned them. At this juncture I was pierced in the left hand with an arrow which passed through the reins and the fourth finger, pinning my hand to the bridle. I instantly broke the spear head and threw it aside. Unmindful of the fact that the feathers could not pass through the wound, I pulled the arrow in the direction in which it had been shot, and was compelled finally in order to free myself of it to seize the feathered in lieu of the barbed end.

Thus raged this hand to hand conflict until all our shots were expended, and it was found that owing to the restiveness of the horses we could not reload while mounted. We then fell back about fifty yards and dismounted for that purpose. Soon afterward arose from beyond the burning heap one continuous mourning howl, such as can alone come forth from the heart of the red man in deep distress. These sounds of sorrow revealed to me that we were in little danger of a renewal of the assault, and I was, I may in truth say, most thankful for the truce thus proclaimed. Two of our men had been killed and four, besides myself, severely wounded; we had also one horse killed and several disabled. Had the combat been renewed I would have had, after leaving a guard with the horses, but five or six men to fight on foot.

Nightfall was approaching; the Indians gathered up their dead and wounded, and moved off toward the Rio Grande. Our thirst, which was great at the beginning of the combat, had now become intense from excitement and loss of blood. I therefore moved at once to Devil's river, where we bivouacked about 10 p. m., and sent a messenger to Camp Hudson for supplies and medical aid.

Thus closed this terrible scene, and often since have I felt most grateful that our horses were so broken down, as but for their condition they would, doubtless, when beaten over the head with shields, have become totally unmanageable, and have caused the massacre of my entire command. I attribute also our escape to the fact that the Indians did not have the self-possession to cut our bridle reins, which act would have proved fatal to us. We were nigh meeting a similar fate to that of the gallant Custer and his noble band.

I learned after the fight, through other Indians as well as through my guide, that the party which attacked us were Comanches and Lipans. The exact number of their killed we were unable to ascertain, owing chiefly to the cover afforded by the Spanish bayonets, but we were confident at the hour that it amounted to not less than nine or ten; we were equally certain that four to one were engaged against us.

Lieutenant Fink came up the following day with a detachment of Infantry. Our troops returned to the scene of action and buried the dead, as I had neither pick nor shovel at the time of the encounter. Moreover I could not have delayed thereafter for any purpose, on account of the extreme suffering of the men for want of water.

After a respite of a few days I marched to Fort Clark and there made a brief report of the affair, which is now, I presume, on file in Washington. General David E. Twiggs, commanding the Department, shortly afterwards published the following order:

"HEADQUARTERS, DEPARTMENT OF TEXAS.
"SAN ANTONIO, *August 5th, 1857.*

"SIR:—Lieutenant Hood's report was transmitted last mail; from subsequent information, not official, I think Lieutenant Hood's estimate of the Indian party was much too small. The same party, it appears, attacked the California mail guard five days after, and near the place where Lieutenant Hood had the fight, and they estimated the Indians to be over one hundred. These affairs were in the vicinity of Camp Hudson where Lieutenant Fink of the Eighth Infantry is stationed with a Company of Infantry. If this company had have been furnished with some fifteen or twenty horses, the second attack would not probably have been made. Lieutenant Hood's affair was a most gallant one, and much credit is due to both the officer and men.

"I am, sir, very respectfully, your obedient servant,

"(Signed) D. E. TWIGGS,
"Brevet Major General, U. S. A., Commanding Department.

"To Lieutenant Colonel L. THOMAS,
"Assistant Adjutant General,
"Headquarters of the Army, West Point, New York."

I also afterwards learned through the Indian Agent that the Indians at the Reservation stated my command had killed nineteen warriors during the fight, and that General Twiggs's estimate was about correct in regard to numbers. The comparatively small loss we sustained is strong evidence that our shots proved most destructive, and that the Indians labored under an intense excitement which caused them generally to miss their mark. The fact that we were mounted and above their level seems to have rendered their aim very imperfect, as shown by the circumstance that one of my wounded men whose horse had been killed, was pierced in the back with three additional arrows (one of which passed through his lung), as he was making his way to the rear of the line.

Early in August I returned to Fort Mason, where not long afterwards I was promoted to the rank of First Lieutenant, assigned to Company "K," and placed on duty at Camp Colorado, on the upper waters of the river of that name. In 1858 I re-established Camp Wood, on the Nueces river, about forty miles distant from its source, and at this post my company continued in the performance of the ordinary duties of soldiers upon the frontier till the declaration of war in 1861.

In November, 1860, I was granted a leave of absence for six months, and on my arrival at Indianola I received an order directing me to report for duty as Chief of Cavalry at West Point. I immediately proceeded to Washington, and made application in person to Colonel S. Cooper, Adjutant General, to be relieved from the order and allowed to avail myself of the leave of absence already granted. I shall ever remember the astonishment of this old and most worthy soldier at my unwillingness to go to West Point. He turned quickly in his chair, saying: "Lieutenant, you surprise me; this is a post and position sought by almost every soldier." I replied it was true, but I feared war would soon be declared between the States, in which event I preferred to be in a situation to act with entire freedom. He acceded to my request; before the expiration of my leave of absence hostilities were inaugurated, and my resignation was tendered to the United States Government.

Shortly before the secession of the Southern States I returned to Camp Wood, and, although still on leave, accompanied my regiment to Indianola, where I bid my comrades a reluctant farewell. Kentucky being the land of my nativity, I deemed it right I should first tender my services for her defence. Accordingly I went to Louisville in the early Spring of 1861, and subsequently to Frankfort. I met the Governor, Mr. Breckinridge and other prominent men of that Commonwealth; but after long debate and considerable delay, I became convinced that no decided action would be taken. I repaired at the latter part of April to Montgomery, Alabama, offered my services to the Confederate Government, was appointed First Lieutenant in the Army and ordered to Richmond to report to Colonel R. E. Lee, who had very recently assumed command of all the troops in Virginia by authority of the Governor of that State.

During my long service in Texas I had had occasion to visit almost ever portion of that extensive and beautiful territory, and was able to form an idea of the future prosperity of that State. So deeply impressed had I become with its vast and undeveloped resources that I had, just prior to the war, determined to resign and make it my home for life. Therefore when Kentucky failed to act, I entered the Confederate service from the State of Texas, which thenceforth became my adopted land. I arrived in Richmond about the 5th of May, sent my luggage to the hotel, and proceeded without delay to the office of Colonel Lee who had, I was informed, been recently promoted to the rank of Major General. He extended me a most cordial greeting, and, taking me by the hand, said: "I am glad to see you. I want you to help me." I replied that I came to Richmond with that object, and inquired what duties he desired me to perform. He answered: "I wish you to go to Yorktown and report to Colonel Magruder." I naturally asked at what time he desired me to leave. He turned his head, looked at the clock, and, with a smile, said he would like me to go before I dined. It was then about 11 a. m., and I well knew he meant early dinner.

I went immediately to the Ballard House, ordered my trunk to the station, and left for Yorktown. On the train I could but contrast the surroundings of General Lee, as I had just beheld him, with the quiet and peaceful scenes we had passed through together but a year or two before upon the frontier of Texas. His office was in the third or fourth story of, I think, the Mechanics' Institute; and he had around him, it seemed to me, every cobbler in Richmond, giving them instructions as to the manner of making cartridge boxes, haversacks, bayonet scabbards, &c. He was studiously applying his great mind to this apparently trivial but most important work. The Confederacy was destitute of such equipments at that hour, and it may be safely asserted that his labor in this regard and in the organization of our troops was the source, in a great measure, of the success of our arms in the engagements which soon followed.

I arrived at Yorktown that afternoon about an hour before sunset, and reported to Colonel Magruder, whom, if I remember correctly, I found out upon the line of works around the town. He forthwith placed me in command of several batteries then in position. Upon my right and left, almost as far as the eye could extend, were infantry regiments in line of battle, and, in their front, officers delivering stirring and warlike appeals to the men. As no tent or quarters had been assigned me, I sent for my trunk and sat upon it in the sand a greater portion of the night, gazing intently every few minutes in the direction of Fortress Monroe, in the expectation momentarily of beholding the enemy. The following morning it was ascertained that the Federals were not within thirty miles of this line bristling with bayonets. The excitement therefore soon subsided, and the soldiers returned to their respective bivouacs. Such was my first night of service in the Confederate Army.

Colonel Magruder assigned me to the command of the cavalry companies then at Yorktown, and directed me to drill and discipline them, and at the same time picket his front. These troops were from Virginia and as fine a body of men as that State sent to the war. I was only a First Lieutenant, and the companies were of course under the direction of

2

captains; a question eventually arose in respect to rank, and Magruder, unwilling to await action at Richmond, declared me Captain by his own order. Subsequently discussion arose touching the date of commissions of the Captains, and he at once, by the same process, declared me, Major. This settled all matters pertaining to authority, and I continued on outpost service, covering the front of Magruder's forces.

Soon after the affair at Big Bethel, it became the custom of the enemy to send out every few days scouting parties of infantry in the direction of our position at Yorktown. I determined to go at night into the swamp lying between the James and York River roads, remain quietly under cover, and, upon the advance of such a party, to move out upon its rear, and capture it if possible. In accordance with this plan, I concealed my troops in the swamp several nights, when finally a battalion of infantry came forth upon the James River road. I moved out in the rear of the Federals, overtook and attacked them upon the same spot where Colonel Dreux, of Louisiana, had been killed. Our assault in rear produced great consternation, and the enemy ran in all directions through the woods. However, we killed several of their number, and captured some ten or fifteen prisoners whom we sent to Yorktown, where the infantry climbed to the house and tree tops to see the first "boys in blue" I presume many of them had ever beheld.

Through orders from Richmond, these cavalry companies were then organized into a regiment. Colonel Robert Johnson was placed in command, and I was promoted to the rank of Lieutenant Colonel. In this position I served until, I think, in July, when I was summoned to Richmond, appointed Colonel, and directed to organize the Fourth Texas Infantry Regiment from the detached companies which had recently arrived from that State, and were at the time in camp near that city. I remained there drilling this splendid body of young men and educating them in the duties of soldiers till September, when we were ordered to join the right of General

Joseph E. Johnston's Army at Dumfries. Honorable L. I. Wigfall had been appointed Brigadier General and assigned to the command of the Texas brigade.

Quarters were constructed by placing the tents on pickets with a chimney attached, which provision made the men comparatively comfortable for the Winter. I remained on the Lower Occoquan during the Winter of 1861-62, engaged in the instruction of my regiment in all its essential duties. I lost no opportunity whenever the officers or men came to my quarters, or whenever I chanced to be in conversation with them, to arouse their pride, to impress upon them that no regiment in that Army should ever be allowed to go forth upon the battle-field and return with more trophies of war than the Fourth Texas;—that the number of colors and guns captured, and prisoners taken, constituted the true test of the work done by any command in an engagement. Moreover, their conduct in camp should be such as not to require punishment, and, when thrown near or within towns, should one of their comrades be led to commit some breach of military discipline, they should, themselves, take him in charge, and not allow his misconduct to bring discredit upon the regiment; proper deportment was obligatory upon them at home, and, consequently, I should exact the same of them whilst in the Army. By perseverance in this system I experienced no difficulty in their management. One of the main obstacles to the attainment of strict discipline, in the training of volunteers, is the issuance of orders without satisfactory explanation as to their object. For example, the usual and important regulation, prohibiting lights or noise in quarters after ten o'clock at night, would be regarded by young recruits as unnecessary, and even arbitrary, unless the officer in command illustrated to them the necessity thereof, and made them understand that an Army in time of active operations must have sleep at night, in order to march and fight the following day; and that for this reason no soldier should be allowed to keep awake, say, six of his comrades in the same tent, nor be permitted to create a disturbance, which would deprive his neighbors of rest, and render them unfit for duty the ensuing morning.

On the 7th of March, 1862, I followed up the movement with my regiment back in the direction of Fredericksburg; en route, and, greatly to my surprise, I received information of my appointment as Brigadier General, and of my assignment to the command of the Texas brigade. General Wigfall, if I remember correctly, had been elected to the Senate, and regarded his services more important in that chamber than upon the field. This promotion occasioned me some annoyance, as Colonel Archer, who commanded the Fifth Texas, and to whom I was warmly attached, ranked me by seniority. He, however, came to my tent, spontaneously congratulated me upon my advancement, and expressed his entire willingness to serve under me. He gave proof of the sincerity of his professions by a subsequent application to be transferred to my division, after I was appointed Major General, and he was promoted to the rank of Brigadier. Moreover, some years later, when I assumed the direction of the Army of Tennessee, he applied for orders to report to me for duty. He was not only a fine soldier, but a man of sterling qualities, and whose nobility of character was unsurpassed.

I had been stationed a few weeks in the vicinity of Fredericksburg, when orders were received to march to Yorktown, at which place we arrived a few days prior to the 17th of April, the date of General Johnston's assumption of the command of all the forces on the Peninsula. I was here placed in reserve with my brigade, which consisted of the First, Fourth, Fifth Texas, and Eighteenth Georgia Regiments, and continued the system of instruction and training already indicated. I had so effectually aroused the pride of this splendid body of men, as to entertain little fear in regard to their action on the field of battle.

The 3d of May, "on information that the Federal batteries would be ready for service in a day or two," the Commanding General ordered the Army to retreat. Accordingly, I marched with my brigade, which formed part of Major General G. W. Smith's Division, upon the Yorktown road, in the direction of Williamsburg. At daybreak of the 5th the retreat was continued from Williamsburg towards Richmond, through deep mud, and in a heavy rain. Whilst in bivouac opposite West Point, General Whiting informed me that a large body of the enemy had disembarked at Eltham's Landing; that our cavalry was on picket upon the high ground overlooking the valley of York river, and instructed me to move my brigade in that direction, and drive the enemy back if he attempted to advance from under cover of his gunboats. Pursuant to imperative orders, the men had not been allowed to march with loaded arms during the retreat. On the 7th, at the head of my command, I proceeded in the direction of Eltham's, with the intention to halt and load the muskets upon our arrival at the cavalry outpost. I soon reached the rear of a small cabin upon the crest of the hill, where I found one of our cavalrymen half asleep. The head of the column, marching by the right flank, with the Fourth Texas in the front, was not more than twenty or thirty paces in my rear, when, simultaneously with my arrival at the station of this cavalry picket, a skirmish line, supported by a large body of the enemy, met me face to face. The slope from the cabin toward the York river was abrupt, and, consequently, I did not discover the Federals till we were almost close enough to shake hands. I leaped from my horse, ran to the head of my column, then about fifteen paces in rear, gave the command, forward into line, and ordered the men to load. The Federals immediately opened fire, but halted as they perceived our long line in rear. Meanwhile, a corporal of the enemy drew down his musket upon me as I stood in front of my line. John Deal, a private in Company "A," Fourth Texas Regiment, and who now resides in Gonzales, Texas, had fortunately, in this instance, but contrary to orders, charged his rifle before leaving camp; he instantly killed the corporal, who fell within a few feet of me. At the time I ordered the leading regiment to change front forward on the first company, I also sent directions to the troops in rear to follow up the movement and load their arms, which was promptly executed. The brigade then gallantly advanced, and drove the Federals, within the space of about two hours, a distance of one mile and a half to the cover of their gunboats. When we struck their main line quite a spirited engagement took place, which, however, proved to be only a temporary stand before attaining the immediate shelter of their vessels of war. Hampton's brigade, near the close of the action, came to our support, and performed efficient service on the right.

Our loss was slight, whereas that of the enemy was quite severe. General Johnston states in his Narrative that if Northern publications of that period are to be relied upon, it was ten times greater than our own. The Commanding General of the Army, though correct in his assertion that the security of his march required the dislodgement of the enemy from its position south of Eltham's Landing, is in error in regard to the troops who bore the brunt of the combat, as will be seen by the following extract from the official report of Major General G. W. Smith, who at that time commanded the division:

> "Referring to the reports of the several commanders for details, it is only necessary for me to state that the Texas brigade, under command of Brigadier General John B. Hood, supported on the right by the Hampton Legion and the Nineteenth Georgia Regiment, of Colonel Hampton's brigade, were selected, and ordered forward by General Whiting, to drive the enemy from the woods then occupied in front of their landing. Late in the day the Tennessee brigade, commanded by Brigadier General Anderson, was placed in position to support and cover the left flank of the Texans. All the troops engaged showed the finest spirit, were under perfect control, and behaved admirably. The brunt of the contest was borne by the Texans, and to them is due the largest share of the honors of the day at Eltham. The Texas brigade lost eight killed and twenty-eight wounded; in the other portions of the command there were twelve wounded and none killed."

This affair, which brought the brigade so suddenly and unexpectedly under fire for the first time, served as a happy introduction to the enemy.

The ensuing day the march was resumed to the rear and continued till we reached the Baltimore Cross-roads, in which vicinity we bivouacked about five days; thence we retreated to a point near Richmond. About this juncture it was rumored that the Commanding General contemplated the abandonment of the Capital of the Confederacy. General McClellan, however, soon threw across the Chickahominy, to the south bank, about one-fourth of his forces, and the Confederate Army was ordered to make ready to assail this detachment. Major General G. W. Smith massed his division on the Nine Miles road the morning of the 31st of May. Longstreet and Hill assembled on the right, lower down on the Chickahominy; they attacked and were driving the enemy handsomely, when about 3 p. m. General Smith ordered General Whiting to advance through the swamp. The object was to assault, on his right flank, the enemy engaged against Longstreet. Law's brigade came in contact with the Federals, as my troops would soon have done, had not General Johnston, in person, unfortunately changed my direction by ordering me to move off by the right flank, and join Longstreet's left. Shortly after I passed the railroad, a battery, to my surprise, fired upon us from the rear. I nevertheless continued to march by the flank; a few moments later, I heard roar upon roar of musketry in the direction of the ground I had just left, and naturally supposed our troops were firing into each other, by mistake. The undergrowth in the swamp through which we were passing was very dense, and the water waist deep in some places; consequently, our progress was not as rapid as I desired. Soon after this heavy firing in rear, Major S. D. Lee came to me in great haste with instructions to return forthwith, as our troops on the left required support, and, at the same time, informed me that General Johnston had been wounded. I immediately started back, but nightfall approached before I was enabled to rejoin Major General Smith, and render him the assistance I would have gladly afforded. The following day my brigade remained in line of battle without encountering the enemy; with this marching and counter-marching ended the part taken by my troops in the battle of Seven Pines or Fair Oaks.

CHAPTER II.

CONFEDERATE STATES ARMY, VIRGINIA—GAINES'S MILLS OR FIRST COLD HARBOR, MALVERN HILL, SECOND MANASSAS, BOONSBORO, GAP, AND SHARPSBURG, OR ANTIETAM.

After the battle of Seven Pines, General R. E. Lee was assigned to the command of the Army of Northern Virginia. He immediately commenced to form plans by which to free the Confederate Capital from the proximity of the enemy. His first move was to send General Whiting's Division to Staunton, as a ruse, to join General Jackson; to order the latter then to march toward Richmond, or down the north side of the Chickahominy, upon the right flank of McClellan; and, when Jackson was sufficiently near the enemy, to throw across this stream the main body of the Confederate Army at, and in the vicinity of Meadow bridge, and, finally, with his united forces to make a general assault upon the Federals. I happened to have been made cognizant of the foregoing plan through General Whiting, just prior to or during the march to Staunton. I mention the source from which I obtained this information, as it might seem strange that a Brigadier General should have knowledge of the secret purposes of such a movement, in operations of so great importance.

My brigade having been reinforced by Hampton's Legion, under the command of Colonel Geary, moved by railway about the middle of June, via Lynchburg, to Charlottesville, and thence marched to Staunton. Upon our arrival at this place, we received orders to retrace our steps, return to Charlottesville, and there take the train to Hanover Junction. On the 25th I conducted my command, which now formed a part of Jackson's Army, to Ashland. At this point rations and ammunition were issued to the troops, and, the morning of the 26th, I marched with my brigade in a southeasterly direction towards Cold Harbor, as the advanced guard of Jackson's forces. We soon came in contact with the Federal outposts, whom we drove rapidly to and across Tottapotamoi creek, a sluggish stream, with banks steep and densely wooded on either side. Here I discovered the bridge on fire, and the enemy busily engaged felling trees to check our advance beyond; thereupon, Reiley's battery was placed in position, and opened fire, whilst we continued to push forward our skirmish line. The Federals finally retreated in such haste that they left their axes in the trees. The bridge was promptly repaired, and we continued skirmishing with their rear guard till we reached Handley's Corner, where we halted, and bivouacked for the night.

We had heard during the day, in the direction of Mechanicsville, the guns of Longstreet and A. P. Hill, which indicated that the issue of the great battle, then in progress, would soon be decided. At early dawn of the 27th the march was resumed; Ewell's Division bore off in the direction of our left during the day, and Whiting's to the right. The latter received instructions, in the afternoon, to repair to the support of Longstreet, then assaulting the Federal left at Cold Harbor. I moved on with all possible speed, through field and forest, in the direction of the firing, and arrived, about 4.30 p. m., at a point, on the telegraph road, I should think not far distant from the centre of our attacking force. Here I found General Lee, seated upon his horse. He rode forward to meet me, and, extending his usual greeting, announced to me that our troops had been fighting gallantly, but had not succeeded in dislodging the enemy; he added, "This must be done. Can you break his line?" I replied that I would try. I immediately formed my brigade in line of battle with Hampton's Legion on the left. In front was a dense woods and ugly marsh, which totally concealed the enemy from us; but the terrible roar of artillery and musketry plainly revealed, however, that thousands and thousands of living souls were struggling in most deadly conflict for the mastery of that field, and I might say, almost under the shadow of the Capitol of the infant Confederacy. My line was established, and moved forward, regiment by regiment, when I discovered, as the disposition of the Eighteenth Georgia was completed, an open field a little to its right. Holding in reserve the Fourth Texas, I ordered the advance, and galloped into the open field or pasture, from which point I could see, at a distance of about eight hundred yards, the position of the Federals. They were heavily entrenched upon the side of an elevated ridge running a little west and south, and extending to the vicinity of the Chickahominy. At the foot of the slope ran Powhite creek, which stream, together with the abatis in front of their works, constituted a formidable obstruction to our approach, whilst batteries, supported by masses of infantry, crowned the crest of the hill in rear, and long range guns were posted upon the south side of the Chickahominy, in readiness to enfilade our advancing columns. The ground from which I made these observations was, however, open the entire distance to their entrenchments. In a moment I determined to advance from that point, to make a strenuous effort to pierce the enemy's fortifications, and, if possible, put him to flight. I therefore marched the Fourth Texas by the right flank into this open field, halted and dressed the line whilst under fire of the long-range guns, and gave positive instructions that no man should fire until I gave the order; for I knew full well that if the men were allowed to fire, they would halt to load, break the allignment, and, very likely, never reach the breastworks. I moreover ordered them not only to keep together, but also in line, and announced to them that I would lead them in the charge. Forward march was sounded, and we moved at a rapid, but not at a double-quick pace. Meantime, my regiments on the left had advanced some distance to the front through the wood and swamp.

Onward we marched under a constantly increasing shower of shot and shell, whilst to our right could be seen some of our troops making their way to the rear, and others lying down beneath a galling fire. Our ranks were thinned at almost every step forward, and proportionately to the growing fury of the storm of projectiles. Soon we attained the crest of the bald ridge within about one hundred and fifty yards of the breastworks. Here was concentrated upon us, from batteries in front and flank, a fire of shell and canister, which ploughed through our ranks with deadly effect. Already the gallant Colonel Marshall, together with many other brave men, had fallen victims in this bloody onset. At a quickened pace we continued to advance, without firing a shot, down the slope, over a body of our soldiers lying on the ground, to and across Powhite creek, when, amid the fearful roar of musketry and artillery, I gave the order to fix bayonets and charge. With a ringing shout we dashed up the steep hill, through the abatis, and over the breastworks, upon the very heads of the enemy. The Federals, panic-stricken, rushed precipitately to the rear upon the infantry in support of the artillery; suddenly the whole joined in the flight toward the valley beyond. At this juncture some twenty guns, stationed in rear of the Federal line on a hill to my left, opened fire upon the Fourth Texas, which changed front, and charged in their direction. I halted in an orchard beyond the works, and despatched every officer of my staff to the main portion of the brigade in the wood on the left, instructing them to bear the glad tidings that the Fourth Texas had pierced the enemy's line, and were moving in his rear, and to deliver orders to push forward with utmost haste. At the same moment I discovered a Federal brigade marching up the slope from the valley beyond, evidently with the purpose to re-establish the line. I ran back to the entrenchments, appealed to some of our troops, who, by this time, had advanced to the breastworks, to come forward and drive off this small body of Federals. They remained, however, motionless. Jenkins's command, if I mistake not, which was further to our right, boldly advanced and put this brigade to rout. Meantime, the long line of blue and steel to the right and left wavered, and, finally, gave way, as the Eighteenth Georgia, the First and Fifth Texas, and Hampton's Legion gallantly moved forward from right to left, thus completing a grand left wheel of the brigade into the very heart of the enemy. Simultaneously with this movement burst forth a tumultuous shout of victory, which was taken up along the whole Confederate line.

I mounted my horse, rode forward, and found the Fourth Texas and Eighteenth Georgia had captured fourteen pieces of artillery, whilst the Fifth Texas had charge of a Federal regiment which had surrendered to it. Many were the deeds of valor upon that memorable field.

General Jackson, in reference to this onset, says in his official report:

> "In this charge in which upwards of a thousand men fell, killed and wounded, before the fire of the enemy, and in which fourteen pieces of artillery and nearly a regiment were captured, the Fourth Texas, under the lead of General Hood, was the first to pierce these strongholds and seize the guns. Although swept from their defences by this rapid and almost matchless display of daring and desperate valor, the well disciplined Federals continued in retreat to fight with stubborn resistance."

On the following day, as he surveyed the ground over which my brave men charged, he rendered them a just tribute when he exclaimed: "The men who carried this position were soldiers indeed!"

Major Warwick, of the Fourth Texas, a brave and efficient officer, fell mortally wounded near the works, whilst urging his men forward to the charge; over one-half of this regiment lay dead or wounded along a distance of one mile. Major Haskell, son-in-law of General Hampton, won my admiration by his indomitable courage: just after my troops had broken the adversary's line, and I was sorely in need of staff officers, he reported to me for duty, sword in hand, notwithstanding one of his arms had by a shot been completely severed from his body. I naturally instructed him to go in search of a surgeon.

After the capture of the artillery posted on the hill in rear of the Federal line, a strange and interesting incident occurred. The Second Cavalry, my regiment in the United States service prior to the war, gallantly charged the Fourth Texas, the regiment I had organized and commanded in the Confederate Army. Major Whiting, who was captain of my company on the frontier of Texas, commanded the former in this bold attack to recapture these guns; his horse was killed under him, and he fell stunned, though unharmed, at the feet of my men, and was taken prisoner.

When the battle had ceased, I gave my attention at once and during the night—to the care of the wounded, as doctors, litter-bearers and ambulance drivers were without much experience at that early period of hostilities. As I rode over the field, about 2 o'clock in the morning, amid the wounded whose touching appeals for water resounded on every side, a voice in the distance arose, calling me by my surname in tones of deep distress. Shortly after one of my soldiers came and reported to me that Captain Chambliss, an old friend, and a member of the Second Cavalry, United States Army, was lying upon the hill, desperately wounded. I ordered him to return immediately, to render every assistance in his power, and to assure Chambliss that I would soon be with him, as I was then completing the necessary arrangements for the care of the wounded. About daybreak I reached the spot where my friend lay, and we met with the same warmth of feeling which had characterized our intercourse previous to the war. I issued instructions to have him transported to the hospital, and accorded the same attention given to my own wounded officers. Although I feared at the time his wounds would prove mortal, he, I am glad to state, finally recovered.

Subsequent to the battles around Richmond, I, in company with Colonel Fitzhugh Lee, also formerly of the Second Cavalry, United States Army, visited the Capital, and, at the Libby prison, called upon Whiting and Chambliss, with whom we renewed the cordial relations we had enjoyed upon the frontier.

The dead were buried on the field of Cold Harbor or Gaines's Mills on the 28th, and, the afternoon of the 29th, my brigade began the pursuit of the enemy along with Jackson's forces. We crossed the Chickahominy at Grapevine bridge, near the railroad; arrived at Savage's Station the morning of the 30th, and pushed on to White Oak Swamp, where we found the enemy in position on the opposite side of the stream, in our immediate front, behind the bridge, which they had destroyed on the retreat. General Jackson ordered forward a few batteries, opened fire, and, at the same time, sent detachments to the right and left to effect a crossing and assail our adversary upon both flanks. Whilst this artillery duel in our front was progressing, Longstreet and A. P. Hill were heavily engaged lower down at Frayser's Farm. At a very early hour on the morning of July 1st we forced the passage of White Oak Swamp, moved rapidly forward, and, before long, reached the field which Hill and Longstreet had compelled the enemy to abandon. From this point Jackson's Corps led the advance of Lee's Army upon the Willis Church road; my brigade, under an annoying fire from the Federal rear guard, soon arrived in an open field in front of and commanded by Malvern Hill. The latter was not only a position of immense natural strength, but was, moreover, crowned with artillery which was supported by McClellan's entire Army.

General Whiting's Division, in this meadow, constituted the left of the Confederate line; and, although the position occupied by the enemy in our immediate front was seemingly impregnable, the country on their right appeared to be open, and to afford an easy approach. I therefore dispatched some of my Texas scouts to reconnoitre in that direction. The report, shortly received, was of a favorable character, and General Hampton and I requested of General Whiting permission to turn and assail this exposed flank. Our application was not granted, however, and we remained during the day under a murderous fire of artillery, whilst our forces on the right were driven back in every attempt made to gain possession of Malvern Hill. The ensuing night the Federals retreated to Harrison's Landing, on the James river, and thus put an end to this bloody and fruitless contest.

General Jackson marched, after this engagement, in the direction of Culpepper Court House, leaving my brigade with Longstreet. The battle at Cedar Run soon followed, and resulted in a brilliant victory for Jackson over Pope, whilst Longstreet remained with his corps in observation of McClellan's shattered forces at Harrison's Landing. A fleet of vessels, however, appeared on the James river to transport the Federals to another field of operations, and orders were issued to march to the Rapidan in the vicinity of Gordonsville, which point we reached about the 15th of August.

My command had been increased by the addition of two or more batteries and a splendid brigade, under Colonel E. M. Law, an able and efficient officer. General Evans was shortly afterwards given, besides his own troops, command of the two brigades under my direction.

On the 20th of August my division, acting as an advanced guard of Longstreet's Corps, moved against General Pope's Army, then lying a short distance south of the Rappahannock, crossed the Rapidan at Raccoon Ford, and marched in the wake of Jackson's Corps, which was pushing forward rapidly with the design to secure a position on the flank or in rear of the Federals. This manœuvre resulted in one of those bold and dazzling achievements which not only won my unbounded admiration, but deservedly earned for Jackson the highest appreciation and encomiums of the civilized world. Whilst he was hastening forward with a determination to allow no obstacle to hinder the accomplishment of his object, his train was attacked by the enemy near Welford's Ford, on Hazel river; nevertheless, true to the inspiration of his genius, he pushed onward, leaving Trimble's brigade to protect his baggage against this assault. General Trimble gallantly repulsed the Federals, as my division moved forward to his support.

Longstreet's Corps continued to threaten the enemy, while Jackson turned his right flank and cut his communications with Washington. He finally stood at bay near Manassas, whilst Longstreet, by a forced march from the Rappahannock, pushed forward, and reached about mid-day, on the 28th, Thoroughfare Gap, which was guarded by a strong force of the enemy.

My command had marched nearly the whole previous night. About 2 a. m., after passing through a valley amid darkness which was greatly increased by a dense wood, the troops were allowed to file off, stack arms, and bivouac on a slope, and around a knoll upon which some of our cavalrymen had been stationed on picket duty. The fatigue of the men was so excessive that they dropped down in line, and fell asleep almost the instant they touched the ground. Amid the stillness and darkness which reigned in the encampment, some of the officers, who had dismounted upon the summit of the hillock, kicked over an empty barrel which had been used by the cavalrymen as a receptacle for forage, and it came rolling and bounding down the slope over the bushes, toward the Texans who were then in a sound sleep. Just at this moment a favorite animal of one of the regiments, "the old grey mare," loaded with kettles, tin cups and frying pans, dashed up the hill from the forest below with a rattling noise. Some one gave the alarm, crying with a loud voice, "Look out!" and the brave men who had fought so nobly at Cold Harbor sprang to their feet, deserted their colors and guns, and ran down the slope over a well-constructed fence, which was soon levelled to the ground, and had continued their flight several hundred yards before they awoke sufficiently to recover their wits, and boldly march back, convulsed with laughter. This incident is the origin of the brigade song, the burden of which ran, "The old Grey Marè came tearing out o' the Wilderness." The truth is, in time of war, a cap explodes much louder at night than in the day.

Late in the afternoon of the 28th my division was instructed to unite with General D. R. Jones's Division and gain possession of Thoroughfare Gap, a narrow mountain defile, protected, as it were, by a wall of stone on either side. At the same time General Lee sent a force to the left to threaten the Federals in rear, whilst a portion of my command passed through the Gap under a heavy fire of artillery, and my main force crossed over the ridge upon the immediate left of the Gap. The enemy was thus forced to retire, and my division bivouacked for the night beyond this stronghold. At early dawn on the morning of the 29th I put my troops in motion, and, in accordance with instructions from General Longstreet, formed his advanced guard in the direction of Manassas. I placed Lieutenant Colonel Upton, of the Fifth Texas, in command of about one hundred and fifty picked men, from the Texas brigade, to act as skirmishers, and instructed him to rapidly push the Federals in his front. I impressed upon him the importance of hastening to the support of General Jackson, and assured him I would keep the division in readiness to render him prompt assistance, if requisite. Here was achieved by this advanced guard of the advanced guard one of those military feats which is entitled to the admiration of every soldier. Although the Federals opposed us with the different arms of the service, Colonel Upton drove them before him with such rapidity that General Longstreet sent me orders, two or three times, to halt, since the Army was unable to keep within supporting distance of my forces. The gallant Upton was, indeed, pre-eminent in his sphere as an outpost officer.

I joined General Jackson on the Groveton pike, upon the field of Manassas, about 10.30 a. m., when he rode forward and extended me a hearty welcome. He was then keeping at bay the entire Federal Army, commanded by Major General

Pope. My division was formed without delay across the pike; the Texas brigade was posted on the right, and that of Law on the left. Between my left and Jackson's right, which rested about one mile south of Groveton, a gap of a few hundred yards existed; it was afterwards filled by artillery, under the direction of Colonel Walton. Longstreet's Corps, as it arrived upon the field, formed on my right, thus constituting my division the centre of the Confederate Army. I was instructed to obey the orders of either Lee, Jackson, or Longstreet. We remained, till a late hour in the afternoon, spectators of the heavy engagement of Jackson's troops with the enemy, who was thwarted in his attempt to turn our left flank. Major B. W. Frobel, whom I had previously assigned to the command of my artillery, was sent to our right with his battalion to oppose a column of the enemy, advancing to attack Longstreet whilst he was establishing his line. He speedily repulsed the Federals, and returned to his former position.*

In the meantime our opponents had been massing their forces in our front. Just before sunset I received orders from General Longstreet to advance, and scarcely had I given the word of command, when the enemy moved forward and began a general attack along my line. Law's brigade of Alabamians, Mississippians and Carolinians dashed forward with the Texans, Georgians and Geary's Legion, upon their immediate right; each seemed to vie with the other in efforts to plunge the deeper into the ranks of the enemy. Onward they charged, driving the foe through field and forest, from position after position, till long after darkness had closed in upon the scene of conflict. Law had captured one piece of artillery,† and I beheld with pride the work done by my men, who had forced back the Federals a distance of over one mile.

I now discovered that my line was in the midst of the enemy; the obscurity of the night, which was deepened by a thick wood, made it almost impossible to distinguish friend from foe,

* See Frobel's Report.

† See Law's Report.

and for the same reason I was unable to select a position and form upon it for action next morning. The Confederates and Federals were so intermingled that commanders of both armies gave orders for allignment, in some instances, to the troops of their opponents. Colonel Work, of the First Texas, was struck in the head with an inverted musket in the hands of a Federal, and several stands of colors were snatched from their bearers by my troops, and borne off as mementos of this night encounter of clubs and fists.

In view of this condition of affairs I determined to ride to the rear, inform Generals Lee and Longstreet of the facts, and to recommend that I retire and resume the line from which I had advanced just before sunset. I found them about two miles off, in an open field, and, after a brief interview, we received orders to act in accordance with my suggestion. The troops were therefore withdrawn from the immediate presence of the enemy, back to their original position across the Groveton pike, about 2 a. m. on the 30th of August. As I was prepared to lie down and rest for the few remaining hours before dawn, one of my officers informed me that General Richard Anderson's Division was bivouacked in mass just in my front. Knowing that some thirty or forty pieces of artillery bore directly upon his troops, I mounted my horse, rode off in search of his quarters, and urged him to hasten his withdrawal, as the Federal artillery would assuredly, at daylight, open upon his men thus massed, and greatly cripple his division. Anderson had been marching all day, in order to join General Lee, and did not halt until he found himself in the midst of Federal and Confederate wounded. Upon my warning, he promptly aroused his men and, just after daybreak, marched to the rear of my line of battle. The pike was dry, and his division, as it moved back, left a cloud of dust in its wake, which circumstance, I have always thought, induced General Pope to send his celebrated despatch to Washington to the effect that General Lee was in full retreat.

My troops remained stationary a greater part of the 30th, quietly awaiting orders to again advance. About 3.30 p. m. a furious assault was made upon Jackson, within full view of my position. Line after line was hurled against his brave men, posted in a railroad cut, from which they stubbornly resisted every attack. I sent for a battery (Reiley's, if I mistake not), and ordered it to open upon the flank of the enemy's attacking column, whilst Colonel S. D. Lee's artillery, together with the remainder of Major Frobel's batteries, ploughed deep furrows through the Federal masses, as they advanced to and recoiled before the "Stonewall" upon my left. So desperate was the assault of Pope, and so fixed the determination of this commander, or some of his officers, to force the troops to fight that a line was, apparently, stationed in rear to fire upon those who, impelled by fear or despair, sought refuge from the battle-field.

Thus raged this fierce contest, when about 4 p. m. I received an order, through one of Longstreet's staff officers, to advance. A few minutes after my division moved forward, a messenger from Longstreet summoned me, and, at the full speed of my horse, I joined him from a quarter to a half of a mile in rear. He instructed me not to allow my division to move so far forward as to throw itself beyond the prompt support of the troops he had ordered to the front. Notwithstanding I rode at as rapid a course as my favorite horse could bear me to rejoin my two brigades, I did not overtake them till I had crossed the creek, about four hundred yards south of the Chinn House, and the Texas brigade had captured a battery, routed the Federal Zouaves—literally strewing the ground with their dead and wounded—and Law, upon the left, had accomplished equally important results in his front. The field, where lay the dead and dying zouaves in their gay uniforms, amid the tall green grass, presented indeed a singular appearance, as I passed down the slope and across the creek. I here sent orders to my troops to halt and adjust their allignment, and discovered, at the same time, upon a ridge a short distance beyond, another battery together with large masses of Federal infantry in the vicinity of the Chinn House. Soon, Colonel Means, mounted and in command of General Evans's brigade, reported to me for directions. I instructed him to take the battery which was then within sixty yards of us. His men boldly dashed forward, and he, a few moments later, fell dead to the ground pierced by a ball.

I moved a little to the right, and about this juncture D. R. Jones's Division arrived upon the scene of action; it was soon followed by the remainder of Longstreet's Corps. General Jones rode up to me, and desired to know at which point he could most effectually strike the enemy. I recommended that he at once assail the heavy lines in rear of the Chinn House. He promptly accepted the suggestion, in concert with several other commanders, and they moved to the attack, as did the whole line from right to left. Thus the splendid corps of Longstreet moved forward in a grand charge out upon the high and open ground in that vicinity. Onward it swept toward Bull Run, driving the enemy at a rapid pace before it, and presenting to the view the most beautiful battle scene I have ever beheld. I was in conference, near the Chinn House, with General Jones and other commanders, as they arrived upon the field, when the Fifth Texas—after Colonel Robertson had been wounded in the faithful discharge of his duty, and the gallant, noble Upton had been killed—*slipped the bridle* and rushed forward, breaking loose from its brigade. When night approached, and the battle was over, I found it far to the front, in the vicinity of the Sudley Ford road.

Whilst I lost many valuable officers and men, as shown by the official reports, my two brigades, true to their teaching, captured five guns in addition to fourteen stands of colors, which they bore off as trophies of war and proof of the noble work they had accomplished. During this engagement Major W. H. Sellers, my Adjutant General, led the Texas brigade. I had ordered him to assume direction when General Longstreet sent for me at the beginning of the movement forward.

This distinguished soldier not only deserves great credit for his conduct in this battle, but proved himself, as I expressed my conviction in my official report, competent to command a brigade at that early period of the war. Toward the close of the battle I pushed forward some of my reliable Texas scouts, and captured a number of new Federal ambulances, with a view to better the outfit of my troops. After nightfall I reassembled my division, and rode back to the headquarters of General Lee.

I found him in an open field, near a camp-fire of boards kindled for the purpose of reading despatches; he was in high spirits, doubtless on account of the brilliant and complete victory just achieved by his Army. He met me in his usual manner, and asked what had become of the enemy. I replied that our forces had driven him almost at a double-quick, to and across Bull Run, and that it was a beautiful sight to see our little battle-flags dancing after the Federals, as they ran in full retreat. He instantly exclaimed, "God forbid I should ever live to see our colors moving in the opposite direction!"

The ambulances I had captured were destined to cause me somewhat of annoyance, which I had nowise anticipated at the time I assigned them to my troops for the use of their sick and wounded. After the burial of the dead on the following day, and the march had been resumed, with orders to follow Jackson's Corps in the direction of Maryland, I was instructed by Major General Evans to turn over these ambulances to his Carolina troops. Whereas I would cheerfully have obeyed directions to deliver them to General Lee's Quarter Master for the use of the Army, I did not consider it just that I should be required to yield them to another brigade of the division, which was in no manner entitled to them. I regarded the command, which had captured them, as the rightful owners in this instance, and therefore refused to obey the order. I was, in consequence, placed in arrest, and, on the march to Frederick, Maryland, was ordered by General Longstreet to proceed to the rear to Culpepper Court House, if I remember correctly, and there await the assembly of a Court Martial for my trial. General Lee, however, became apprised of the matter, and at once sent instructions that I should remain with my command, though he did not release me from arrest. Longstreet's Corps was finally massed near Hagerstown, and by this time my division had become restive and somewhat inclined to insubordination on account of my suspension. I repressed all demonstrations of feeling by assurances to the officers that the affair would soon be settled, and I shortly restored to command.

On the 13th of September intelligence was received of McClellan's advance from the direction of Federal City toward South Mountain, and on the morning of the 14th I marched with Longstreet's Corps to Boonsboro' Gap, a narrow and winding pass, through which runs the turnpike from Hagerstown to Federal City. I was still under arrest, with orders to move in rear of my two brigades. The division reached the foot of South Mountain about 3.30 p. m., from which point could be seen the shells of the enemy, as they passed över the rugged peaks in front, and burst upon the slope in our proximity. I could hear the men, as they filed up the ascent, cry out along the line, "Give us Hood!" but did not comprehend the meaning of this appeal till I arrived with the rear of the column at the base of the ridge, where I found General Lee standing by the fence, very near the pike, in company with his chief of staff, Colonel Chilton. The latter accosted me, bearing a message from the General, that he desired to speak to me. I dismounted, and soon stood in his presence, when he said: "General, here I am just upon the eve of entering into battle, and with one of my best officers under arrest. If you will merely say that you regret this occurrence, I will release you and restore you to the command of your division." I replied, "I am unable to do so, since I cannot admit or see the justness of General Evans's demand for the ambulances my men have captured. Had I been ordered to turn them over for the general use of the Army, I would cheerfully have acquiesced." He again urged me to make some declaration expressive of regret. I answered that I could not consistently do so. Then, in a voice betraying the feeling which warmed the heart of this noble and great warrior, he said, "Well, I will suspend your arrest till the impending battle is decided."

I quickly remounted, galloped to the front of my column, and, with a kind welcome from my troops, reported for duty to General Longstreet, who by this time had reached the summit of the mountain. He immediately instructed me to file to the left, in the wake of Evans's brigade, and to take position with my right near the pike. The advance of McClellan's long lines could be seen moving up the slope in our front, evidently with the purpose to dislodge our forces posted upon the sharp ridge overlooking the valley below. Before long Major Fairfax, of Longstreet's staff, came to me in haste with orders to move to the right of the pike, as our troops on that part of the field had been driven back. He accompanied me to the pike, and here turned his horse to leave, when I naturally asked if he would not guide me. He replied, " No, I can only say, go to the right." Meantime Major Frobel's batteries had come forward into position on top of the ridge; they opened fire, and performed excellent service in checking the enemy. The wood and undergrowth were dense, and nothing but a pig path seemed to lead in the direction in which I was ordered. Nevertheless, I conducted my troops obliquely by the right flank, and while I advanced I could hear the shouts of the Federals, as they swept down the mountain upon our side. I then bore still more obliquely to the right, with a view to get as far as possible towards the left flank of the enemy before we came in contact. We marched on through the wood as rapidly as the obstacles in our passage would admit. Each step forward brought nearer and nearer to us the heavy Federal lines, as they advanced, cheering over their success and the possession of our dead and wounded. Finally, I gave instructions to General Law and Colonel Wofford, directing the two brigades, to order their men to fix bayonets; and, when the enemy came within seventy-five or a hundred yards, I ordered the men to front and charge. They obeyed promptly, with a genuine Confederate yell, and the Federals were driven back pell mell, over and beyond the mountain, at a much quicker pace than they had descended. Night closed in with not only our dead and wounded, together with those of our adversary in our possession, but with the mountain, on the right, within our lines.

After the correction of my allignment I rode, at about 10 p. m., back to the Gap, where I found General D. H. Hill and other officers on the gallery of a tavern, near the pike, evidently discussing the outlook. As I approached, I inquired, in an ordinary tone of voice, as to the condition of affairs on our left, and to my surprise was met with a mysterious "Pshe—Pshe"—; a voice added in an audible whisper, "The enemy is just there in the corn field; he has forced us back." I thereupon suggested that we repair without delay to General Lee's headquarters, and report the situation. Accordingly, we rode down to the foot of the mountain, where we found General Lee in council with General Longstreet. After a long debate, it was decided to retire and fall back towards Sharpsburg.

The morning of the 15th our forces were again in motion in the direction of the Antietam; the cavalry and my two brigades, in addition to Major Frobel's artillery, formed the rear guard to hold our opponents in check, whilst the Army marched quietly to its destination. My troops, at this period, were sorely in need of shoes, clothing and food. We had had issued to us no meat for several days, and little or no bread; the men had been forced to subsist principally on green corn and green apples. Nevertheless, they were in high spirits and defiant, as we contended with the advanced guard of McClellan the 15th and forenoon of the 16th. During the afternoon of this day I was ordered, after great fatigue and hunger endured by my soldiers, to take position near the

Hagerstown pike, in an open field in front of * Dunkard Church. General Hooker's Corps crossed the Antietam, swung round with its right on the pike, and, about an hour before sunset, encountered my division. I had stationed one or two batteries upon a hillock, in a meadow, near the edge of a corn field and just by the pike. The Texas brigade had been disposed on the left, and that of Law on the right. We opened fire, and a spirited action ensued, which lasted till a late hour in the night. When the firing had in a great measure ceased, we were so close to the enemy that we could distinctly hear him massing his heavy bodies in our immediate front.

The extreme suffering of my troops for want of food induced me to ride back to General Lee, and request him to send two or more brigades to our relief, at least for the night, in order that the soldiers might have a chance to cook their meagre rations. He said that he would cheerfully do so, but he knew of no command which could be spared for the purpose; he, however, suggested I should see General Jackson and endeavor to obtain assistance from him. After riding a long time in search of the latter, I finally discovered him alone, lying upon the ground, asleep by the root of a tree. I aroused him and made known the half-starved condition of my troops; he immediately ordered Lawton's, Trimble's and Hays's brigades to our relief. He exacted of me, however, a promise that I would come to the support of these forces the moment I was called upon. I quickly rode off in search of my wagons, that the men might prepare and cook their flour, as we were still without meat; unfortunately the night was then far advanced, and, although every effort was made amid the darkness to get the wagons forward, dawn of the morning of the 17th broke upon us before many of the men had had time to do more than prepare the dough. Soon thereafter an officer of Lawton's staff dashed up to me, saying, "General Lawton sends his compliments with the request that you come

* In my official report erroneously called St. Mumma Church.

at once to his support." "To arms" was instantly sounded, and quite a large number of my brave soldiers were again obliged to march to the front, leaving their uncooked rations in camp.

Still, indomitable amid every trial, they moved off by the right flank to occupy the same position we had left the night previous. As we passed, about sunrise, across the pike and through the gap in the fence just in front of Dunkard Church, General Lawton, who had been wounded, was borne to the rear upon a litter, and the only Confederate troops, left on that part of the field, were some forty men who had rallied round the gallant Harry Hays. I rode up to the latter, and, finding that his soldiers had expended all their ammunition, I suggested to him to retire, to replenish his cartridge boxes, and reassemble his command.

The following extract from the official report of General Jackson will convey an idea of the bloody conflict in which my two little brigades were about to engage:

> "General Lawton, commanding division, and Colonel Walker, commanding brigade, were severely wounded. More than half of the brigades of Lawton and Hays were either killed or wounded, and more than a third of Trimble's, and all the regimental commanders in those brigades, except two, were killed or wounded. Thinned in their ranks, and exhausted of their ammunition, Jackson's Division and the brigades of Lawton, Trimble and Hays retired to the rear, and Hood, of Longstreet's command, again took the position from which he had been before relieved."

Not far distant in our front were drawn up, in close array, heavy columns of Federal infantry; not less than two corps were in sight to oppose my small command, numbering, approximately, two thousand effectives. However, with the trusty Law on my right, in the edge of the wood, and the gallant Colonel Wofford in command of the Texas brigade on the left, near the pike, we moved forward to the assault. Notwithstanding the overwhelming odds of over ten to one against us, we drove the enemy from the wood and corn field back upon his reserves, and forced him to abandon his guns on our left. This most deadly combat raged till our last round of ammunition was expended. The First Texas Regiment had lost, in the corn field, fully two-thirds of its number; and whole ranks of brave men, whose deeds were unrecorded save in the hearts of loved ones at home, were mowed down in heaps to the right and left. Never before was I so continuously troubled with fear that my horse would further injure some wounded fellow soldier, lying helpless upon the ground. Our right flank, during this short, but seemingly long, space of time, was toward the main line of the Federals, and, after several ineffectual efforts to procure reinforcements and our last shot had been fired, I ordered my troops back to Dunkard Church, for the same reason which had previously compelled Lawton, Hays and Trimble to retire.

My command remained near the church, with empty cartridge boxes, holding aloft their colors whilst Frobel's batteries rendered most effective service in position further to the right, where nearly all the guns of the battalion were disabled. Upon the arrival of McLaws's Division, we marched to the rear, renewed our supply of ammunition, and returned to our position in the wood, near the church, which ground we held till a late hour in the afternoon, when we moved somewhat further to the right and bivouacked for the night. With the close of this bloody day ceased the hardest fought battle of the war.

In the Military Biography of Stonewall Jackson, edited by Rev. J. Wm. Jones, D. D., occur the following passages (pp. 330-31) in reference to this engagement:

"Seeing Hood in their path the enemy paused, and a Northern correspondent writes: 'While our advance rather faltered, the rebels, greatly reinforced, made a sudden and impetuous onset,* and drove our gallant fellows back over a portion of the hard won field. What we had

---

* The above mentioned large reinforcements were my two small brigades.

won, however, was not relinquished without a desperate struggle, and here, up the hills and down, through the woods and the standing corn, over the ploughed land and clover, the line of fire swept to and fro as one side or the other gained a temporary advantage.'

"Hood was now fighting with his right toward the main line of the enemy, for General Hooker had swept round so far, that, as we have said, his line was almost at right angles with its original position. Hood threw himself into the action with great gallantry, and says in his report: 'Here I witnessed the most terrible clash of arms by far that has occurred during the war. The two little giant brigades of my command wrestled with the mighty force, and although they lost hundreds of their officers and men, they drove them from their position, and forced them to abandon their guns on our left.' 'One of these brigades numbered only eight hundred and fifty-four (854) men.' "

The following morning I arose before dawn and rode to the front where, just after daybreak, General Jackson came pacing up on his horse, and instantly asked, "Hood, have they gone?" When I answered in the negative, he replied "I hoped they had," and then passed on to look after his brave but greatly exhausted command.

The subjoined letter, I have no doubt, obtained my promotion about this period. I had no knowledge of its existence until after the close of the war, when it was handed to me in New York by Mr. Meyer, to whom I am indebted for the favor. He was at the time of the surrender a clerk in the War Office, at Richmond, and, in consideration of the unsettled condition of affairs, placed it among his papers for preservation:

"HEADQUARTERS, V. DIST.,
"*Sept. 27th, 1862.*

"GENERAL:—I respectfully recommend that Brig. Genl. J. B. Hood be promoted to the rank of a Major General. He was under my command during the engagements along the Chickahominy, commencing on the 27th of June last, when he rendered distinguished service. Though not of my command in the recently hard fought battle near Sharpsburg, Maryland, yet for a portion of the day I had occasion to give directions respecting his operations, and it gives me pleasure to say that his duties were discharged with such ability and zeal, as to command my admiration. I regard him as one of the most promising officers of the army.

"I am, General, your obedient servant,

"(Signed) T. J. JACKSON,

"Major General."

"General S. COOPER,

"Adjutant and Inspector General, C. S. A.

"Endorsed, New York, November 9th, 1866."

"The enclosed letter from General Jackson to General Cooper was handed to General Hood by Mr. Meyer (a former clerk in the War Department at Richmond), at the Southern Hotel in this city. The letter is the original, and preserved by Mr. Meyer.

"(Signed) F. S. STOCKDALE."

The foregoing letter is doubly kind in its tenor, inasmuch as I was not serving in General Jackson's Corps at the time.

During the 18th the Confederate Army remained in possession of the field, buried the dead, and that night crossed near Shepherdstown to the south side of the Potomac. Soon thereafter my division marched to a point north of Winchester, and passed a pleasant month in the beautiful Valley of the Shenandoah. My arrest, which General Lee, just prior to the battle of Boonsboro' Gap, had been gracious enough to suspend, was never reconsidered; the temporary release became permanent, and, in lieu of being summoned to a Court Martial, I was shortly afterwards promoted to the rank of Major General with the command of two additional brigades.

The accession of Benning's and Anderson's brigades, which had already taken part in a number of battles, composed a division which any general might justly have felt honored to command. The former brigade had been gallantly led by General Toombs at Sharpsburg. I experienced much interest in training these troops, as I endeavored to excite emulation among them and thoroughly arouse their pride, in accordance with the system of education I had pursued with the Fourth Texas Regiment, Law's, and my original brigade. Under the unfortunate organization of brigades by States, I lost the Eighteenth Georgia Regiment and Hampton's Legion, to both of which commands, I, as well as my Texas troops, had become warmly attached. The former had served with me longer than the latter, and in every emergency had proved itself bold and trusty; it styled itself, from a feeling of brotherhood, the Third Texas.

Whilst I lost these two excellent bodies of men, I gained the Third Arkansas, a large regiment, commanded by Colonel Van Manning, a brave and accomplished soldier, who served with distinction, and, in truth, merited higher rank and a larger command. I also lost the Sixth North Carolina, Ninth and Eleventh Mississippi Regiments, which, after long and gallant service in Law's brigade, were also transferred to other commands; thus, unfortunately, were severed relations which had been engendered and strengthened by common trials and dangers.

CHAPTER III.

CONFEDERATE STATES ARMY—VIRGINIA—FREDERICKSBURG, SUFFOLK, GETTYSBURG, AND CHICKAMAUGA.

THE latter part of October McClellan's movements determined General Lee to withdraw from the Valley of the Shenandoah, leaving his cavalry in rear, and to return to the Valley of the Rappahannock. Accordingly, my division took its place, about the 26th, in the marching columns of Longstreet's Corps, which moved in the direction of the latter point. During the previous month of quiet and rest, the troops had received a supply of shoes and clothing, and had improved in drill and discipline. This splendid corps, therefore, exhibited a very different appearance from that which it presented in its ragged and bare-footed condition, a short period before in Maryland.

We halted in the vicinity of Culpepper Court House, where shortly afterwards intelligence was received that McClellan had been superseded by the appointment of Burnside. This General promptly made a demonstration on the Upper Rappahannock, as he moved towards Fredericksburg. General Lee crossed to the south side of the Rapidan, and, by the latter part of November, the Federal and Confederate Armies again confronted one another at Fredericksburg, where we quietly awaited the development of events.

On the 11th of January, 1863, General Burnside having completed all necessary preparation, began to lay pontoons above and below the railroad bridge which had been destroyed. That entire day and night he consumed in crossing his forces to the southern bank of the river, under cover of, at least, one hundred pieces of artillery. During the 12th he formed his line below and above Deep Run, whilst upon the range of hills overlooking the valley, Lee's forces lay in readiness to receive the attack. General Jackson had, meantime, moved up to form line on our right, and that day, if I remember correctly, as we were riding together in direction of General Lee's headquarters, the conversation turned upon the future, and he asked me if I expected to live to see the end of the war. I replied that I did not know, but was inclined to think I would survive; at the same time, I considered it most likely I would be badly shattered before the termination of the struggle. I naturally addressed him the same question, and, without hesitation, he answered that he did not expect to live through to the close of the contest. Moreover, that he could not say that he desired to do so. With this sad turn in the conversation, the subject dropped. Often since have I thought upon these words, spoken casually by each of us, and which seem to have contained the prophecy of his untimely death and of my own fate.

My division was again the centre of the Confederate Army, as it rested in line of battle opposite Deep Run, full of spirit and impatient for action. The following morning, after the fog had disappeared, and at about 10 o'clock, the heavy lines of the enemy advanced upon our right and against Jackson's forces, but were driven back beneath the fire of our guns posted on that part of the line. Again, at about 1 p. m., the attack was renewed, and the Federals penetrated into a gap left in Jackson's front line. They were, however, speedily repulsed by his brigades held in reserve. My troops repelled with ease the feeble attack made on their immediate front, whilst

Longstreet's remaining forces on the left drove the enemy back repeatedly with great slaughter near Marye's Hill.

I was directed in this battle, as at Second Manassas, to obey the orders either of Generals Lee, Jackson, or Longstreet. About sunset, after the musketry fire had nigh ceased, I received instructions through an officer of Jackson's staff to join in a movement on my right as soon as A. P. Hill's division advanced. The order was accompanied with a message from General Jackson that he intended to drive the enemy into the river. I responded that I was in readiness to act, but, for some reason unknown to me, these orders were countermanded.

About 10 o'clock that night I rode back to my encampment to procure a cup of coffee, and, General Lee's quarters being within a few hundred yards, I walked up the ridge and presented myself at his tent. He immediately asked me what I thought of the attack by the enemy during the day. I expressed my opinion that Burnside was whipped; that no good general would ever make an assault similar to that upon my right and left, without intending it as his main effort, and that the heavy roll of musketry I had heard clearly convinced me that the hardest part of the battle had been fought. He then remarked that he did not think Burnside had made his principal attempt, but would attack again the next day, and that we would drive him back and follow him up to the river. After conversing a few moments longer, during which time he was in the highest spirits, I returned to my line, where I continued the remainder of the night.

The morning of the 14th both Armies still lay face to face, no aggressive movement having been initiated by either side, when about noon Generals Lee and Jackson rode by my position, and invited me to accompany them on a reconnoissance towards our right. We soon reached an eminence, not far distant from Hamilton's Crossing on the railroad, and upon which some of our batteries were posted. From this point we had a magnificent view of the Federal lines on their left, some seven in number, and each, seemingly, a mile in length.

General Jackson here turned to me, and asked my estimate of the strength of the enemy then in sight and in our immediate front. I answered fifty thousand, and he remarked that he had estimated their numbers at fifty-five thousand.

Strange to say, amid this immense assemblage of Federal troops not a standard was to be seen; the colors were all lowered, which circumstance induced me to abide by the opinion I had expressed to General Lee the night previous. The two Armies stood still during this entire day, and the following morning we awoke to find the enemy on the north side of the Rappahannock.

In this vicinity my division was quartered for the Winter, and my tent remained near that of General Lee. It was my privilege to often visit him during his leisure hours, and converse with the freedom of yore upon the frontier. In one of our agreeable chats, in company with General Chilton, his chief of staff, he complained of his Army for burning fence rails, killing pigs, and committing sundry delinquencies of this character. I spoke up warmly in defence of my division, declaring that it was not guilty of these misdemeanors, and desired him to send Chilton to inspect the fences in the neighborhood of my troops. General Lee, who was walking up and down near his camp fire, turned toward me and laughingly said, "Ah, General Hood, when you Texans come about the chickens have to roost mighty high." His raillery excited great merriment, and I felt I was somewhat at a stand; nevertheless, I urged that General Chilton be sent at least to inspect the fences.

Time passed pleasantly till the early Spring, when General Longstreet marched back to Petersburg, and thence towards Suffolk—a movement I never could satisfactorily account for, and which proved unfortunate, since it allowed General Hooker, who had superseded Burnside the latter part of April, to cross the Rappahannock and attack General Lee in the absence of one-half of his Army. The transcendent genius of "Stonewall," by which he executed one of his most brilliant moves to the rear of the assailants, once more thwarted the Federal Commander, who was hurled back beyond the Rappahannock to seek refuge upon Stafford Heights. But alas! at a terrible sacrifice, an irreparable loss to the Confederacy: the immortal Jackson.

I had received information of Hooker's anticipated advance, and was most anxious to rejoin my old chief, General Lee. Never did I so long to be with him as in this instance, and I even proceeded so far as to apply for permission to move with my division to his support. The request, however, was not granted.

Longstreet, after receiving the order to join General Lee, made every effort to accomplish this great end, but his wagons were, unfortunately, out in search of forage, and the march was consequently delayed; for which reason we failed to reach Chancellorsville in time to participate in the battle.

Nothing was achieved against the enemy on the expedition to Suffolk, at which point he possessed a safe place of refuge within his strong fortifications, protected by an impenetrable abatis. During our sojourn in this vicinity, quite a spirited affair occurred between our troops and the Federal gunboats, on the Nansemond river, and in which I suffered a grave misfortune in the loss of Captain Turner, of the Fifth Texas. As an outpost officer, he was gifted with the same pre-eminent qualities which distinguished the gallant Upton.

On the march from Suffolk to Chancellorsville, intelligence reached us of the Confederate victory and of the death of Jackson. This latter event occasioned me deep distress. I was hereupon prompted to write to General Lee, giving expression to my sorrow, and, at the same time, to my regret at our failure to join him before the great battle he had just fought and won. In reply to my brief note, he addressed me as follows:

"CAMP FREDS, *21st May, 1863.*

"MY DEAR GENERAL:—Upon my return from Richmond, I found your letter of the 13th awaiting me. Although separated from me, I have always had you in my eye and thoughts. I wished for you much in the last battle, and believe had I had the whole Army with me, General Hooker would have been demolished. But God ordered otherwise.

"I grieve much over the death of General Jackson—for our sakes, not for his. He is happy and at peace. But his spirit lives with us, and I hope it will raise up many Jacksons in our ranks. We must all do more than formerly. We must endeavor to follow the unselfish, devoted, intrepid course he pursued, and we shall be strengthened rather than weakened by his loss. I rely much upon you. You must so inspire and lead your brave division, as that it may accomplish the work of a corps. I agree with you as to the size of the corps of this Army. They are too large for the country we have to operate in for one man to handle. I saw it all last campaign. I have endeavored to remedy it—this in a measure at least—but do not know whether I shall succeed. I am much obliged to you always for your opinion. I know you give it from pure motives. If I am not always convinced, you must bear with me. I agree with you also in believing that our Army would be invincible if it could be properly organized and officered. There never were such men in an Army before. They will go anywhere and do anything if properly led. But there is the difficulty—proper commanders—where can they be obtained? But they are improving—constantly improving. Rome was not built in a day, nor can we expect miracles in our favor.

"Wishing you every health and happiness, and committing you to the care of a kind Providence,

"I am now and always your friend,

"(Signed) R. E. LEE.

"General J. B. Hood,
"Commanding Division."

Again early in May we were in bivouac in the Rapidan, and preparations were initiated for another campaign. The artillery and transportation were carefully inspected, and whatever was found unserviceable was sent to the rear. At this period my division was in splendid condition, its four brigades being under the direction of Law, Benning, Anderson and Robertson. Past service had created with each command a feeling of perfect confidence in its associate whenever brought under fire. The artillery had again been increased by the addition of a number of pieces, as will be seen by the following report of Colonel Owen:

"HEADQUARTERS BATTALION WASHINGTON ARTILLERY,
"NEW ORLEANS, *February 15th, 1879.*

"*Copy of Report of Major* HENRY'S *Battalion of Artillery, July 19th, 1863, attached to* HOOD'S *Division, First (Longstreet's) Corps, Army of Northern Virginia:*

BATTERY COMMANDERS.	12 Napoleons.	10 Parrots.	3 inch Rifle.
Captain Buckman,	4		
Captain Garden,	3	1	
Captain Reiley,	2	3	1
Captain Latham,	2	2	
	11	6	1

"Official copy from original return, 18.
"(Signed) W. M. OWEN,
"Late Adjutant to Chief Artillery First Corps."

This battalion completed the organization of as brave and heroic a division, numbering, approximately, eight thousand effectives, as was ever made ready for active service. So high-wrought was the pride and self-reliance of the troops that they believed they could carve their way through almost any number of the enemy's lines, formed in the open field in their front.

Soon after the 1st of June the Confederate forces crossed the Rapidan, and advanced again in the direction of Maryland. About the middle of the month we forded the Potomac, which was so swollen by recent rain that the men were forced to uplift their cartridge boxes, in order to keep dry their ammunition. Nevertheless, they marched in regular order to the northern bank of that beautiful stream, and, as they moved through the deep water the inspiriting strains of "Dixie" burst forth from bands of music. Never before, nor since, have I witnessed such intense enthusiasm as that which prevailed throughout the entire Confederate Army.

Shortly afterwards we crossed into Pennsylvania, amid extravagant cheers which re-echoed all along the line. Our forces marched undisturbed, and were massed in the vicinity of Chambersburg, where intelligence was received of General Meade's assignment to the command of the Federal Army.

My headquarters were again in close proximity to those of General Lee, and, after a few days devoted to rest and quiet, I, as usual, rode to pay him my respects. I found him in the same buoyant spirits which pervaded his magnificent army. After the ordinary salutation, he exclaimed, "Ah! General, the enemy is a long time finding us; if he does not succeed soon, we must go in search of him." I assured him I was never so well prepared or more willing.

A few days thereafter, we were ordered to Gettysburg, and to march with all possible speed.

The following letter, which I addressed General Longstreet in 1875, gives, up to the hour I was wounded and borne from the field, an account of the part taken by my command in the great battle which ensued:

"NEW ORLEANS, LA., *June 28th, 1875.*

"GENERAL JAMES LONGSTREET:—General, I have not responded earlier to your letter of April 5th, by reason of pressure of business, which rendered it difficult for me to give due attention to the subject in regard to which you have desired information.

"You are correct in your assumption that I failed to make a report of the operations of my division around Suffolk, Va., and of its action in the battle of Gettysburg, in consequence of a wound which I received in this engagement. In justice to the brave troops under my command at this period, I should here mention another cause for this apparent neglect of duty on my part. Before I had recovered from the severe wound received at Gettysburg, your corps (excepting Pickett's Division) was ordered to join General Bragg, in the West, for battle against Rosecranz; my old troops—with whom I had served so long—were thus to be sent forth to another Army—quasi, I may say, among strangers—to take part in a great struggle; and upon an appeal from a number of the brigade and regimental officers of my division, I consented to accompany them, although I had but the use of one arm. This movement to the West soon resulted in the battle of Chickamauga, where I was again so seriously wounded as to cause the loss of a limb. These severe wounds in close succession, in addition to the all-absorbing duties and anxieties attending the last year of the war, prevented me from submitting subsequently a report, as likewise one after the battle of Chickamauga, in which engagement—whilst you led the left wing—I had the honor of commanding your corps together with three divisions of the Army of Tennessee, respectively under A. P. Stewart, Bushrod Johnson and Hindman. Thus, the gallantry of these troops, as well as the admirable conduct of my division at Gettysburg, I have left unrecorded.

"With this apology for seeming neglect, I will proceed to give a brief sketch, from memory, of the events forming the subject of your letter:

"My recollection of the circumstances connected with the attempt, whilst we were lying in front of Suffolk, to reach General Lee in time to participate in the battle of Chancellorsville, is very clear. The order directing your corps to move to the support of General Lee, was received about the time Hooker crossed the Rappahannock. Unfortunately we had been compelled by scarcity of forage to send off our wagons into North Carolina to gather a supply from that State. A short delay necessarily ensued, as couriers had to be dispatched for requisite transportation before the troops could move. Every effort, however, was made to get to Lee at the earliest moment. If my memory betrays me not, you repaired in advance of your corps to Petersburg or Richmond, having issued orders for us to march with all possible speed to Lee, on the Rappahannock. I was most anxious to get to the support of my old chief, and made strenuous efforts to do so; but, whilst on a forced march to accomplish this object, I received intelligence of our victory at Chancellorsville, and of Jackson's mortal wound. We, nevertheless, continued our march, and eventually went into bivouac upon the Rapidan, near Gordonsville.

"After the battle of Chancellorsville, preparations were made for an offensive campaign.

"Accordingly, my troops moved out of camp, crossed the Rapidan about the 5th June, 1863, and joined in the general move in the direction of the Potomac. We crossed the river about the middle of the same month, and marched into Pennsylvania. Hill's and Ewell's Corps were in advance, and were reported to be in the vicinity of Carlisle. Whilst lying in camp, not far distant from Chambersburg, information was received that Ewell and Hill were about to come in contact with the enemy near Gettysburg. My troops, together with McLaws's Division, were put in motion upon the most direct road to that point, which, after a hard march, we reached before or at sunrise on the 2d of July. So imperative had been the orders to hasten forward with all possible speed, that on the march my troops were allowed to halt and rest only about two hours, during the night from the 1st to the 2d of July.

"I arrived with my staff in front of the heights of Gettysburg shortly after daybreak, as I have already stated, on the morning of the 2d of

July. My division soon commenced filing into an open field near me, where the troops were allowed to stack arms and rest until further orders. A short distance in advance of this point, and during the early part of that same morning, we were both engaged in company with Generals Lee and A. P. Hill, in observing the position of the Federals. General Lee—with coat buttoned to the throat, sabre-belt buckled round the waist, and field glasses pending at his side—walked up and down in the shade of the large trees near us, halting now and then to observe the enemy. He seemed full of hope, yet, at times, buried in deep thought. Colonel Freemantle, of England, was ensconced in the forks of a tree not far off, with glass in constant use, examining the lofty position of the Federal Army.

"General Lee was, seemingly, anxious you should attack that morning. He remarked to me, 'The enemy is here, and if we do not whip him, he will whip us.' You thought it better to await the arrival of Pickett's Division—at that time still in the rear—in order to make the attack; and you said to me, subsequently, whilst we were seated together near the trunk of a tree: 'The General is a little nervous this morning; he wishes me to attack; I do not wish to do so without Pickett. I never like to go into battle with one boot off.'

"Thus passed the forenoon of that eventful day, when in the afternoon—about 3 o'clock—it was decided to no longer await Pickett's Division, but to proceed to our extreme right and attack up the Emmetsburg road. McLaws moved off, and I followed with my division. In a short time I was ordered to quicken the march of my troops, and to pass to the front of McLaws.

"This movement was accomplished by throwing out an advanced force to tear down fences and clear the way. The instructions I received were to place my division across the Emmetsburg road, form line of battle, and attack. Before reaching this road, however, I had sent forward some of my picked Texas scouts to ascertain the position of the enemy's extreme left flank. They soon reported to me that it rested upon Round Top Mountain; that the country was open, and that I could march through an open woodland pasture around Round Top, and assault the enemy in flank and rear; that their wagon trains were packed in rear of their line, and were badly exposed to our attack in that direction. As soon as I arrived upon the Emmetsburg road, I placed one or two batteries in position and opened fire. A reply from the enemy's guns soon developed his lines. His left rested on or near Round Top, with line bending back and again forward, forming, as it were, a concave line, as approached by the Emmetsburg road. A considerable body of troops was posted in front of their main line, between the Emmetsburg road and Round Top Mountain. This force was in line of battle upon an eminence near a peach orchard.

"I found that in making the attack according to orders, viz.: up the Emmetsburg road, I should have first to encounter and drive off this advanced line of battle; secondly, at the base and along the slope of the mountain, to confront immense boulders of stone, so massed together as to form narrow openings, which would break our ranks and cause the men to scatter whilst climbing up the rocky precipice. I found, moreover, that my division would be exposed to a heavy fire from the main line of the enemy in position on the crest of the high range, of which Round Top was the extreme left, and, by reason of the concavity of the enemy's main line, that we would be subject to a destructive fire in flank and rear, as well as in front; and deemed it almost an impossibility to clamber along the boulders up this steep and rugged mountain, and, under this number of cross fires, put the enemy to flight. I knew that if the feat was accomplished, it must be at a most fearful sacrifice of as brave and gallant soldiers as ever engaged in battle.

"The reconnoissance of my Texas scouts and the development of the Federal lines were effected in a very short space of time; in truth, shorter than I have taken to recall and jot down these facts, although the scenes and events of that day are as clear to my mind as if the great battle had been fought yesterday. I was in possession of these important facts so shortly after reaching the Emmetsburg road, that I considered it my duty to report to you, at once, my opinion that it was unwise to attack up the Emmetsburg road, as ordered, and to urge that you allow me to turn Round Top, and attack the enemy in flank and rear. Accordingly, I despatched a staff officer, bearing to you my request to be allowed to make the proposed movement on account of the above stated reasons. Your reply was quickly received, 'General Lee's orders are to attack up the Emmetsburg road.' I sent another officer to say that I feared nothing could be accomplished by such an attack, and renewed my request to turn Round Top. Again your answer was, 'General Lee's orders are to attack up the Emmetsburg road.' During this interim I had continued the use of the batteries upon the enemy, and had become more and more convinced that the Federal line extended to Round Top, and that I could not reasonably hope to accomplish much by the attack as ordered. In fact, it seemed to me the enemy occupied a position by nature so strong —I may say impregnable—that, independently of their flank fire, they could easily repel our attack by merely throwing and rolling stones down the mountain side, as we approached.

"A third time I despatched one of my staff to explain fully in regard to the situation, and suggest that you had better come and look for yourself. I selected, in this instance, my adjutant-general, Colonel Harry

Sellers, whom you know to be not only an officer of great courage, but also of marked ability. Colonel Sellers returned with the same message, 'General Lee's orders are to attack up the Emmetsburg road.' Almost simultaneously, Colonel Fairfax, of your staff, rode up and repeated the above orders.

"After this urgent protest against entering the battle at Gettysburg, according to instructions—which protest is the first and only one I ever made during my entire military career—I ordered my line to advance and make the assault.

"As my troops were moving forward, you rode up in person; a brief conversation passed between us, during which I again expressed the fears above mentioned, and regret at not being allowed to attack in flank around Round Top. You answered to this effect, 'We must obey the orders of General Lee.' I then rode forward with my line under a heavy fire. In about twenty minutes, after reaching the peach orchard, I was severely wounded in the arm, and borne from the field.

"With this wound terminated my participation in this great battle. As I was borne off on a litter to the rear, I could but experience deep distress of mind and heart at the thought of the inevitable fate of my brave fellow-soldiers, who formed one of the grandest divisions of that world-renowned army; and I shall ever believe that had I been permitted to turn Round Top Mountain, we would not only have gained that position, but have been able finally to rout the enemy.

"I am, respectfully, yours,

"J. B. HOOD."

Notwithstanding the seemingly impregnable character of the enemy's position upon Round Top Mountain, Benning's brigade, in concert with the First Texas Regiment, succeeded in gaining temporary possession of the Federal line; they captured three guns, and sent them to the rear. Unfortunately, the other commands, whose advance up a steep ascent, was impeded by immense boulders and sharp ledges of rock, were unable to keep pace up the mountain side in their front, and render the necessary support. Never did a grander, more heroic division enter into battle; nor did ever troops fight more desperately to overcome the insurmountable difficulties against which they had to contend, as Law, Benning, Anderson and Robertson nobly led their brave men to this unsuccessful assault. General Law, after I was wounded, assumed command of the division, and proved himself, by his courage and ability, fully equal to the responsibilities of the position.

The losses were very heavy, as shown by the reports, and have often caused me the more bitterly to regret that I was not permitted to turn Round Top Mountain.

The following officers of my staff, most of whom served with me throughout the war, rendered gallant and efficient service, not only in this great battle, but upon many fields where we were thrown together in the heat of action: Colonel W. H. Sellers, Assistant Adjutant General; Colonel E. H. Cunningham, Inspector General; Major B. H. Blanton, Captain John Smith, Captain James Hamilton, Lieutenant E. B. Wade, Aides-de-Camp; Major N. B. George, Quarter Master; Major Jonas, Commissary; and Captain D. L. Sublett, Ordnance Officer, faultlessly discharged their duties in their respective departments. Dr. John T. Darby, Chief Surgeon, distinguished himself by his untiring energy in caring for the wounded; the eminent talent which he displayed in his province, during our struggle, has since deservedly won for him a high position in the medical world.

My official reports bear testimony to the valuable services of other gentlemen temporarily attached to my headquarters. In truth, I can say with pride that no General was ever more ably supported by staff officers than myself, during the war.

When the Confederate Army fell back from Gettysburg, I followed our marching column in an ambulance, suffering very much from the wound received in my arm. In the same vehicle lay General Hampton, so badly wounded that he was unable to sit up, whereas I could not lie down. We journeyed together in this manner to Staunton, a distance of some two hundred miles. Along the pike were seen our wounded, making their way to the rear, and the noble women of Virginia, standing by the wayside to supply them with food, and otherwise administer to their wants.

I remained for a period of one month under medical treatment, first at Staunton and then at Charlottesville, whence I proceeded to Richmond. About the 14th of September my division passed through the Capital, under orders to join General Bragg in the West for the purpose of taking part in battle against Rosecranz. Although I had but partially recovered, I determined, for reasons already stated in my letter to General Longstreet, to place my horse upon the train, and follow in their wake.

I arrived at Ringgold, Georgia, on the afternoon of the 18th, and there received an order from General Bragg to proceed on the road to Reid's bridge, and assume command of the column then advancing on the Federals. I had my horse to leap from the train, mounted with one arm in a sling, and, about 3 p. m., joined our forces, then under the direction of General Bushrod Johnson and in line of battle. A small body of Federal cavalry was posted upon an eminence a short distance beyond. On my arrival upon the field I met for the first time after the charge at Gettysburg a portion of my old troops, who received me with a touching welcome. After a few words of greeting exchanged with General Johnson, I assumed command in accordance with the instructions I had received, ordered the line to be broken by filing into the road, sent a few picked men to the front in support of Forrest's Cavalry, and began to drive the enemy at a rapid pace. In a short time we arrived at Reid's bridge across the Chickamauga, and discovered the Federals drawn up in battle array beyond the bridge, which they had partially destroyed. I ordered forward some pieces of artillery, opened fire, and, at the same time, threw out flankers to effect a crossing above and below and join in the attack. Our opponents quickly retreated. We repaired the bridge, and continued to advance till darkness closed in upon us, when we bivouacked in line, near a beautiful residence which had been fired by the enemy, and was then almost burned to the ground. We had driven the Federals back a distance of six or seven miles. Meantime, the main body of the Army crossed the Chickamauga at different points, and concentrated that night in the vicinity of my command.

General Bragg having formed his plan of attack the following morning, I was given, in addition to my own division, the direction of Kershaw's and Johnson's Divisions, with orders to continue the advance. We soon encountered the enemy in strong force, and a heavy engagement ensued. All that day we fought, slowly but steadily gaining ground. Fierce and desperate grew the conflict, as the foe stubbornly yielded before our repeated assaults; we drove him, step by step, a distance of fully one mile, when nightfall brought about a cessation of hostilities, and the men slept upon their arms.

In the evening, according to my custom in Virginia under General Lee, I rode back to Army headquarters to report to the Commander-in-Chief the result of the day upon my part of the line. I there met for the first time several of the principal officers of the Army of Tennessee, and, to my surprise, not one spoke in a sanguine tone regarding the result of the battle in which we were then engaged. I found the gallant Breckinridge, whom I had known from early youth, seated by the root of a tree, with a heavy slouch hat upon his head. When, in the course of brief conversation, I stated that we would rout the enemy the following day, he sprang to his feet, exclaiming, "My dear Hood, I am delighted to hear you say so. You give me renewed hope; God grant it may be so."

After receiving orders from General Bragg to advance the next morning as soon as the troops on my right moved to the attack, I returned to the position occupied by my forces, and camped the remainder of the night with General Buckner, as I had nothing with me save that which I had brought from the train upon my horse. Nor did my men have a single wagon, or even ambulance in which to convey the wounded. They were destitute of almost everything, I might say, except pride, spirit, and forty rounds of ammunition to the man.

During that night, after a hard day's fight by his old and trusty troops, General Longstreet joined the Army. He reported to General Bragg after I had left Army headquarters, and, the next morning, when I had arranged my columns for the attack and was awaiting the signal on the right to advance, he rode up, and joined me. He inquired concerning the formation of my lines, the spirit of our troops, and the effect produced upon the enemy by our assault. I informed him that the feeling of officers and men was never better, that we had driven the enemy fully one mile the day before, and that we would rout him before sunset. This distinguished general instantly responded with that confidence which had so often contributed to his extraordinary success, that we would *of course* whip and drive him from the field. I could but exclaim that I was rejoiced to hear him so express himself, as he was the first general I had met since my arrival who talked of victory.

He was assigned to the direction of the left wing, and placed me in command of five divisions: Kershaw's, A. P. Stewart's, Bushrod Johnson's, and Hindman's, together with my own. The latter formed the centre of my line, with Hindman upon my left, Johnson and Stewart on the right, and Kershaw in reserve. About 9 a. m. the firing on the right commenced; we immediately advanced and engaged the enemy, when followed a terrible roar of musketry from right to left. Onward we moved, nerved with a determination to become masters of that hotly contested field. We wrestled with the resolute foe till about 2.30 p. m., when, from a skirt of timber to our left, a body of Federals rushed down upon the immediate flank and rear of the Texas brigade, which was forced to suddenly change front. Some confusion necessarily arose. I was at the time on my horse, upon a slight ridge about three hundred yards distant, and galloped down the slope, in the midst of the men, who speedily corrected their allignment. At this moment Kershaw's splendid division, led by its gallant commander, came forward, as Hindman advanced to the attack a little further to the left. Kershaw's line formed, as it were, an angle with that of the Federal line, then in full view in an open space near the wood. I rode rapidly to his command, ordered a change of front forward on his right, which was promptly executed under a galling fire. With a shout along my entire front, the Confederates rushed forward, penetrated into the wood, over and beyond the enemy's breastworks, and thus achieved another glorious victory for our arms. About this time I was pierced with a Minie ball in the upper third of the right leg; I turned from my horse upon the side of the crushed limb and fell—strange to say, since I was commanding five divisions—into the arms of some of the troops of my old brigade, which I had directed so long a period, and upon so many fields of battle.

Long and constant service with this noble brigade must prove a sufficient apology for a brief reference, at this juncture, to its extraordinary military record from the hour of its first encounter with the enemy at Eltham's Landing, on York river, in 1862, to the surrender at Appomattox Court House. In almost every battle in Virginia it bore a conspicuous part. It acted as the advanced guard of Jackson when he moved upon McClellan, around Richmond; and, almost without an exceptional instance, it was among the foremost of Longstreet's Corps in an attack or pursuit of the enemy. It was also, as a rule, with the rear guard of the rear guard of this corps, whenever falling back before the adversary. If a ditch was to be leaped, or fortified position to be carried, General Lee knew no better troops upon which to rely. In truth, its signal achievements in the war of secession have never been surpassed in the history of nations.

The members of this heroic band were possessed of a streak of superstition, as in fact I believe all men to be; and it may here prove of interest to cite an instance thereof. I had a favorite roan horse, named by them "Jeff Davis;" whenever he was in condition I rode him in battle, and, remarkable as it may seem, he generally received the bullets and bore me unscathed. In this battle he was severely wounded on Saturday; the following day, I was forced to resort to a valuable mare in my possession, and late in the afternoon was shot from the saddle. At Gettysburg I had been unable to mount him on the field, in consequence of lameness; in this engagement I had also been shot from the saddle. Thus the belief among the men became nigh general that, when mounted on old Jeff, the bullets could not find me. This spirited and fearless animal performed his duty throughout the war, and after which he received tender care from General Jefferson and family of Seguin, Texas, until death, when he was buried with appropriate honors.

When wounded I was borne to the hospital of my old division, where a most difficult operation was performed by Dr. T. G. Richardson, of New Orleans. He was at the time Chief Medical Officer of the Army of Tennessee, and is now* the President of the Medical Association of the United States.

The day after the battle I was carried upon a litter some fifteen miles to the residence of Mr. Little, in Armuchee Valley. I remained there about one month under the attentive care of Mr. and Mrs. Little, the parents of the gallant Colonel Little, of my division, and under the able medical attendance of Dr. John T. Darby.

I then received intelligence from General Bragg that the enemy was contemplating a raid to capture me. I at once moved to Atlanta, and thence to Richmond.

General Longstreet, has since the war, informed me that he telegraphed the authorities of the Confederate Government from the battle field, on the day I was wounded, urging my promotion to the rank of Lieutenant General, and was kind enough about the same time to send the following letter:

"HEADQUARTERS, CHATTANOOGA,
"*September 24th, 1863.*

"GENERAL:—I respectfully recommend Major General J. B. Hood for promotion to the rank of Lieutenant General, for distinguished conduct and ability in the battle of the 20th inst. General Hood

* 1878–79.

handled his troops with the coolness and ability that I have rarely known by any officer, on any field, and had the misfortune, after winning the battle, to lose one of his limbs.

"I remain, sir, very respectfully,

"Your obedient servant,

"(Signed) J. LONGSTREET,

"Lieutenant General."

"General S. COOPER,

"Adjutant and Inspector General.

"Endorsed:

"Headquarters, near Chattanooga, September 24th, 1863."

"W. D. 1988.

"J. Longstreet, Lieutenant General, recommends Major General J. B. Hood for promotion to the rank of Lieutenant General for distinguished services in the battle of the 20th inst."

"I cordially unite in this just tribute.

"BRAXTON BRAGG,

"General."

"Respectfully submitted to the Secretary of War.

"By order ED. A. PALFREY,

"Lieutenant Colonel and Assistant Adjutant General.

"Respectfully submitted to the President."

"I cannot too warmly express my appreciation of the character and services of this distinguished officer, and cordially concur in recommending his promotion, if only as an appropriate testimonial of the gratitude of the Confederacy.

"J. A. SEDDON,

"Secretary of War.

"3d October, 1863."

"The services of Major General Hood, and his character as a soldier and patriot, are equal to any reward, and justify the highest trust. The recommendation to confer additional rank, as a testimonial, must have been hastily made. The law prescribes the conditions on which Lieutenant Generals may be appointed. Please refer to act.

"JEFFERSON DAVIS.

"October 3d, 1863."

The subjoined extract from a letter of the Hon. Mr. Seddon, Secretary of War, addressed to Senator Wigfall will explain the endorsement of President Davis:

"RICHMOND, VA.
"*October 14th, 1863.*

* * * * "I have felt the deepest interest for your friend, and I trust I may say mine, the gallant Hood. He is a true hero, and was the Paladin of the fight. I need not say how willingly I would have manifested my appreciation of his great services and heroic devotion by immediate promotion, and but for some rigid notions the President had of his powers (you know how inflexible he is on such points), he, too, would have been pleased to confer the merited honor." * * *

I remained in Richmond, and, having been blessed, with a good constitution, rapidly recovered from my wound. By the middle of January, 1864, I was again able to mount my horse and enjoy exercise. My restoration was so complete that I was enabled to keep in the saddle when on active duty, and, during the remainder of the war, never to require an ambulance either day or night. Often President Davis was kind enough to invite me to accompany him in his rides around Richmond, and it was thus I was for the first time afforded an opportunity to become well acquainted with this extraordinary man, and illustrious patriot and statesman of the South. His wonderful nerve and ability, displayed at a most trying epoch of our history, commanded my admiration; he was not only battling with enemies abroad, but with a turbulent Congress at home.

It was during our pleasant excursions round Richmond that he imparted to me his purpose to largely re-enforce General J. E. Johnston's Army at Dalton, for the object of moving in the early Spring to the rear of the Federal Army, then concentrating at Chattanooga. He also expressed a desire to send me to command a corps under General Johnston. I was deeply impressed with the importance of this movement, and cheerfully acquiesced in the proposition of the President, but with the understanding that an aggressive campaign would be initiated. I was loth, indeed, to leave General Lee and the troops with whom I had served for so long a period.

I was promoted to the rank of Lieutenant General, left Richmond about the 1st of February, arrived at Dalton,

Georgia, on the 4th, and reported for duty to General J. E. Johnston.

A short time before leaving the Capital General Breckinridge, whilst we were together in my room at the Spotswood Hotel, approached the seat I was occupying, and placed his hands upon my head, saying, "My dear Hood, here you are beloved by your fellow-soldiers, and, although badly shattered, with the comfort of having done noble service, and without trouble or difficulty with any man." In truth, the course of my official duties up to this hour had not, I might say, been ruffled in any degree. My relations with my superiors, as well as with officers of lesser rank, had been of a most friendly character. But alas, after a journey over a smooth sea for many days—aye three years—a storm suddenly arose which lasted not only to the close of the war, but a long period thereafter.

The foregoing chapters, which contain a brief record of my experiences up to the day I reported for duty in the Army of Tennessee, were written after the body of this work was prepared for publication. As the Dalton-Atlanta campaign presents no action which rises to the dignity of a general battle, and since the strictures of General Johnston demand my earnest attention, I shall here discontinue the relation of events in the order which I have thus far observed, and resume the narrative at the period I assumed command of the Army around Atlanta. I shall substitute a reply to the erroneous and injurious statements in my regard, brought forward by General Johnston, and which will sufficiently record the part I bore in the campaign of that Spring and early Summer.

CHAPTER IV.

REPLY TO GENERAL JOHNSTON—EFFECTIVE STRENGTH AND LOSSES, ARMY OF TENNESSEE—DALTON TO ATLANTA.

I VERY much regret I should find it incumbent upon me to discuss, at this hour, certain operations in the West; but most unjust strictures, passed upon me by General Johnston, and which are derogatory to my character, alike as a man and a soldier, compel me to speak in self-defence, or otherwise admit by silence the charges brought forth.

Although I feel by reason of injustice done me in the past that I have good cause to demand of our people the privilege of a hearing upon certain matters little understood by them, I would, nevertheless, have left the work of vindication to the unbiassed historian of the future, had not my words and actions been so strangely misrepresented.

Before and just after the close of the war, our people, in the despair of defeat, were in no state of mind to listen to truth which ran counter to their prejudices. Blind passion, however, has now subsided, and reason, it is hoped, has returned. I therefore solicit a hearing upon the subject of some of the most important historical events recounted by General Johnston, and in which I was a prominent actor.

In his Narrative General Johnston speaks as follows, pages 353-54:

"General Hood in his report of his own disastrous operations accused me of gross official mis-statements of the strength of the Army and of its losses—asserting that I had 'at and near Dalton' an available force of seventy-five thousand (75,000) men, and that twenty-two thousand five hundred (22,500) of them were lost in the campaign, including seven thousand (7000) prisoners. He recklessly appealed for the truth of these assertions to Major Kinloch Falconer, Assistant Adjutant General, by whom the returns of the Army were made, which were my authority for the statement attacked by General Hood. At my request, made in consequence of this attack, Major Falconer made another statement from the data in his possession, which contradicts the appellant. By that statement the effective strength of the Army 'at and near Dalton' was forty thousand four hundred and eighty-four (40,484) infantry and artillery, and twenty-three hundred and ninety (2390) cavalry."

Furthermore, page 356, he says:

"The loss of the Confederate Army in this campaign, while under my command, was nine thousand nine hundred and seventy-two (9972) killed and wounded. About a third of it occurred near Dalton and Resaca."

The point in controversy, and which I shall consider at present, is this statement made in my official report:

"On the 6th of May, 1864, the Army lay at and near Dalton, awaiting the advance of the enemy. Never had so large a Confederate Army been assembled in the West. Seventy thousand (70,000) effective men were in the easy direction of a single commander, whose good fortune it was to be able to give successful battle, and redeem the losses of the past. * * The Army of Tennessee lost twenty-two thousand seven hundred and fifty (22,750) men, nearly one-third of its strength."

I shall now demonstrate the actual loss during General Johnston's campaign. In order to do this, it is necessary to prove what was the force "at and near Dalton," or as I expressed it in my official report, "In the easy direction of a single commander," May 6th, 1864. It must be admitted that in order to estimate the loss of an Army during any campaign, siege, or battle, it is necessary first to ascertain the total effective force at the beginning of hostilities, and when the battle is over, or the siege, or campaign ended, again to find out the effective total, which, subtracted from the number at the outset, will unquestionably give the loss. This is the only means by which it can be fairly indicated. The losses of an Army are greater or less according to the manner in which the troops are handled; *i. e.*, an Army standing its ground and fighting, or advancing and driving the enemy, as was the case in Virginia, under General Lee, will count but few stragglers and deserters; the actual loss is not great, from the fact that the wounded men go to their homes proud of their wounds, and the majority of them are soon found again in the ranks.

On the other hand, an Army fighting and retreating at the same time, taking up positions, day after day, to be given up only under cover of darkness, suffers great loss. During such a campaign, the orders necessary to be issued in withdrawing from the immediate presence of the enemy, are depressing, such as directing that dead silence be observed, wheels muffled, etc., for fear of discovery and being fired upon. Let this policy be continued for a distance of one hundred miles, as it was from Dalton to Atlanta, and the "pride, pomp and circumstance of glorious war" are lost in a somewhat funereal procession.

The wounded cannot return home buoyant and hopeful, as they are forced to bear with them the chilling intelligence that the Army is falling back; in all such instances they tarry with their friends, and many fail to report again for duty. However, the loss from this source is but small in comparison to that which accrues from the number of stragglers picked up by the enemy, and of deserters who, beholding their homes abandoned to the foe, become disheartened, and return to their families within the lines of the enemy, as was the case in North Georgia and West Alabama during General Johnston's continued retreat.

The statement derived from Doctor Foard's* return of the killed and wounded, is doubtless correct; but General Johnston's intention cannot, assuredly, be to affirm that this number, nine thousand nine hundred and seventy-two (9972), constitutes his entire loss during his campaign. According to the return of Major Falconer, his own Adjutant General,† and to which he refers, the effective strength of the Army on the 10th of June, near Kennesaw Mountain, when about eighty miles from Dalton, and within about twenty miles of Atlanta, was fifty-nine thousand two hundred and forty-eight (59,248); whilst the return of the 10th of July shows, just after crossing the Chattahoochee river on the night of the 9th, an effective total of only fifty thousand six hundred and twenty-seven (50,627), which, subtracted from the number we had when near Kennesaw Mountain the 10th of June, demonstrates a loss of eight thousand six hundred and twenty-one (8621), less six hundred (600) of J. K. Jackson's command, sent to Savannah. Therefore, it seems impossible that this General should wish to create the impression that nine thousand nine hundred and seventy-two (9972) was his entire loss from all causes, when, within the last twenty miles of his retreat, he lost eight thousand and twenty-one (8021) effective men.

A vacancy in the ranks, brought about by desertion, is as actual and effective as if the soldier had been killed in battle. It is worse in its results, as the deserter generally takes with him his arms, and demoralizes the comrades he has forsaken.

I shall now pass to New Hope Church, a little higher up the country, and ascertain his effective force, to which must be added three thousand three hundred and eighty-eight (3388)‡ killed and wounded prior to his arrival at that point, or to "the passage of the Etowah," since they were effective soldiers at Dalton; in this manner I shall gradually trace the number of available troops, from which deduct the effective

---

* Johnston's Narrative, page 576. † Johnston's Narrative, page 574.

‡ Johnston's Narrative, page 325.

total turned over to me by General Johnston on the 18th July, and I shall finally arrive at his entire loss during the campaign.

The Army reached New Hope Church on the 25th and 26th of May, and remained in that vicinity about ten days previous to the retreat upon Pine and Kennesaw Mountains, near Marietta. It was here visited by General L. T. Wigfall, a man of talent, and, at that time, in the Confederate States Senate, but who, owing to his intense enmity to President Davis, allowed himself to be governed by undue influences. General Wigfall was virtually the political chief of staff of General Johnston, and considering the close relations of these gentlemen, a statement from him relative to the strength of the Army at that period may safely be regarded as good authority. This Senator, in a speech directed against President Davis and myself, in the Confederate States Senate, asserted that, "at New Hope Church, he (Johnston) had of all arms* sixty-four thousand (64,000); of these eight thousand (8000) were cavalry, supposing it not to have increased by recruiting up to that time; that gives him fifty-six thousand (56,000) infantry and artillery.

Thus he allowed fifty-six thousand (56,000) infantry and artillery on the 26th May, after being out thirteen days from Dalton; but admitted only eight thousand (8000) cavalry. There must be a mistake in respect to this arm of the service. It should be borne in mind that General Johnston reports, in accordance with Major Falconer's statement, on the 1st of May, and previous to General Sherman's advance, only two thousand three hundred and ninety-two (2392) cavalry, and that no other return was made up until the 10th June, when the Army was near Kennesaw Mountain—forty days in the interim having elapsed. Field-returns are made up from the returns of corps commanders, and may be called for every ten days, or every month or two, as the Commanding General may deem

* Johnston's Narrative, page 591.

proper. It does not follow, however, that commanders of corps, divisions, brigades, and regiments, neglect to make up their returns every few days. In fact, it is well known that this duty is neglected by no discreet officer, even during an active campaign; otherwise there would be no means of ascertaining the number of men engaged in any one battle. The return of Major Falconer, I presume, is correct so far as it gives the effective strength of the cavalry directly at Dalton on the 1st of May; but it does not include brigades "near Dalton," "within the easy direction" of General Johnston, as shown by the following extract from the official return now in the possession of Major General Wheeler, a copy of which this officer furnished me on the 2d May, 1874:

"May 6th, 1864. General field and staff and company officers present, five hundred and twenty-five (525); total effective fighting force, four thousand two hundred and ninety-nine (4299); aggregate officers and men effective for battle, four thousand eight hundred and twenty-four (4824). Dibbrell and Harrison joined from East Tennessee with fourteen hundred and fifteen (1415) effective men just after this report was made. Dibbrell and Harrison reached Resaca about May 1st. I went down and inspected the command."

This aggregate gives a total effective of six thousand two hundred and thirty-nine (6239), and it is evident that General Johnston's chief of cavalry, Major General Wheeler, had in his command this number, "at and near Dalton," not only on the 6th, but on the 1st of May.

General Johnston himself furnishes proof of the correctness of Major General Wheeler's report of the 6th, by his acknowledgment of the presence of Dibbrell's brigade on the 9th of May, in these words:*

"On the same day, Major General Wheeler, with Dibbrell's and Allen's brigades, encountered a large body of Federal cavalry near Varnell's Station." He admits also General

* Johnston's Narrative, page 307.

Martin's division of cavalry to have been at Cartersville a short distance south of Resaca on the 1st of May, and Major Kinloch Falconer states in his official report:* "The cavalry of the Mississippi Army which joined near Adairsville was estimated at three thousand nine hundred (3900) effective men, and Martin's cavalry division, which joined near Resaca, at three thousand five hundred (3500)." Let us, therefore, continue the search for cavalry, before returning to New Hope Church to make the first estimate of the effective strength of this Army. General Johnston, in his Narrative, alludes to the following accessions (p. 353): "Jackson's three thousand nine hundred (3900) met us at Adairsville on the 17th." This number, added to Wheeler's and Martin's forces of six thousand two hundred and thirty-nine (6239), gives of this arm of the service an effective total of ten thousand one hundred and thirty-nine (10,139); which number, in lieu of eight thousand (8000) reported at New Hope Church, added to fifty-six thousand (56,000) infantry and artillery, gives sixty-six thousand one hundred and thirty-nine (66,139), instead of sixty-four thousand (64,000), of all arms, as stated by General Wigfall.

The following letter, from Major General Wheeler affords additional evidence of the correctness of the foregoing estimate of cavalry:

"NEW ORLEANS, LA., *June 1st, 1866.*

"DEAR GENERAL:—In reply to your inquiry as to the aggregate number of officers and enlisted men, mounted and dismounted, I could have thrown into action at any time prior to the siege of Atlanta, had I been notified that a battle was to be fought, and time given to bring up men detailed at the rear, I will state that the records in my possession show that upon the 10th day of July, 1864, the figures referred to were eight thousand four hundred and ten (8410); of these, six thousand two hundred and seventy-nine (6279) were fighting, enlisted men already on duty, mounted on serviceable horses. General Jackson had three brigades which are not included in the above.

* Johnston's Narrative, page 574.

"I cannot give his exact force, but when he first joined the Army near Adairsville I was informed he had about forty-five hundred (4500) men.

"I remain, General, with great respect,

"Your obedient servant,

"(Signed) JOSEPH WHEELER.

"To General JOHN B. HOOD,

"Late Commanding Army of Tennessee."

Thus, the first summary shows an effective total of sixty-six thousand one hundred and thirty-nine (66,139) men, thirteen days out from Dalton; to which force should be added three thousand three hundred and eighty-eight (3388) killed and wounded, a loss which General Johnston acknowledges to have sustained prior to the passage of the Etowah, and the result shows a grand Army of sixty-nine thousand five hundred and twenty-seven (69,527) effectives "at and near Dalton," exclusive of deserters, stragglers, and the prisoners captured from the commencement of the campaign to the arrival of the Army at New Hope Church. Therefore it requires an allowance of but four hundred and seventy-three (473) men, lost as stragglers and deserters during the thirteen days of retreat, in order to sum up the seventy thousand (70,000) effectives alluded to in my official report.

I shall now ascertain the strength of the Army at Adairsville, on the 17th May, after four days' retreat, and again estimate the effective strength "at or near Dalton."

Shortly after the fall of Atlanta, and whilst we were lying in bivouac at Lovejoy Station, I sent for Major Kinloch Falconer, who was at that time one of my Assistant Adjutant Generals, and called his attention to the outcry against me, through the medium of the press, which charged that I had lost many more men during the siege of Atlanta than had General Johnston during his campaign; and, inasmuch as he was the Adjutant General of my predecessor, I desired to know from him the entire loss, from all causes, during the retreat from Dalton to Atlanta. He at once replied that he could not give me the exact figures, for the reason that General Johnston had taken with him all the books and records of the Army to Macon, but that we had lost, in round numbers, twenty-five thousand (25,000) men. Moreover, that we had, at Adairsville, fifty-three thousand (53,000) infantry. Two of my staff officers, Captain John Smith and Lieutenant E. B. Wade, happened to be present at the time, and gave me, whilst these facts were fresh in their memory, the following affidavits:

"HOUSTON, TEXAS, *June 21st, 1865.*

"I certify that on or about the 10th September, 1864, Major Kinloch Falconer, Assistant Adjutant General, Army of Tennessee, reported officially in the presence of Lieutenant E. B. Wade, Aide-de-Camp, Mr. James H. Haggerty, and myself, to General J. B. Hood, at Lovejoy's Station, Ga., that the loss of that Army, from all causes during the campaign from Dalton to Atlanta, Ga., between the 7th of May and 18th of July, 1864, was twenty-five thousand (25,000) effective men. He also stated that the Army, when at Adairsville, Ga., numbered fifty-three thousand (53,000) effective infantry, after the losses sustained at Rocky Face Mountain and Resaca, Ga.

"(Signed) JOHN SMITH,
"Aide-de-Camp.

"Sworn to and subscribed before me this 22d June, A. D. 1865.

"(Signed) WILLIAM ANDREWS,
"Mayor of Houston."

"I certify that on or about the 10th day of September, 1864, Major Kinloch Falconer, late Assistant Adjutant General, Army of Tennessee, C. S. A., reported officially in the presence of Captain John S. Smith, Aide-de-Camp, Mr. Haggerty, and myself, to General J. B. Hood, commanding Army at Lovejoy Station, Ga., that the loss of that Army, from all causes, during the campaign from Dalton to Atlanta, Ga., between the 7th day of May and the 18th day of July, 1864, was twenty-five thousand (25,000) effective men. He (Major Kinloch Falconer) also stated that the Army when at Adairsville, Ga., numbered fifty-three thousand (53,000) effective infantry, after the losses sustained at Rocky Face Mountain and Resaca, Ga.

"(Signed) E. B. WADE,
"Aide-de-Camp.

"STATE OF TENNESSEE,
"RUTHERFORD COUNTY. }

"This day came before me, J. N. Clark, J. P. for said county, E. B. Wade, and made oath that the facts stated in the within certificate are true to the best of his knowledge and belief, this 1st day of June, 1866.

"E. B. WADE.

"Sworn and subscribed before me the date above.

"(Signed) J. N. CLARK, J. P.
"For said County."

The statement of Major Falconer relative to the strength of the infantry at Adairsville tallies very well with that of General Wigfall, as to this arm of the service at New Hope Church, and I have no doubt of the correctness of Major Falconer's assertion to me.

Allowing three thousand eight hundred (3800) artillery, acknowledged by Major Falconer on the 10th of June, the Army at Adairsville exhibits, with the addition of ten thousand one hundred and thirty-nine (10,139) cavalry, an effective total of sixty-six thousand nine hundred and thirty-nine (66,939), to which should be added three thousand three hundred and eighty-eight (3388) killed and wounded "near Dalton and Resaca," and this second summary shows an effective total of seventy thousand three hundred and twenty-seven (70,327), exclusive of stragglers, deserters, and prisoners captured from Dalton to Adairsville.

The foregoing summary fully tallies with the first, and we find, when only four days out from Dalton, over seventy thousand effectives, exclusive of deserters, stragglers, and prisoners captured.

It cannot be asserted with any degree of reason that it was not feasible to have remained longer at Dalton, inasmuch as General Johnston had only to fortify Mill Creek and Snake Creek Gaps to insure his safety, and sufficient time to receive all the reinforcements then en route to Dalton.

As further evidence of the correctness of my assertion, this General states in his Narrative, page 352:

"The troops received by the Army of Tennessee during the campaign were those sent and brought to it by Lieutenant General Polk, and formed the corps of the Army which he commanded. Of these, Canty's Division of about three thousand (3000) effectives reached Resaca on the 9th of May. Loring's of five thousand (5000) on the 11th; French's of four thousand (4000) joined us at Cassville on the 18th; and Quarles's brigade of twenty-two hundred (2200) at New Hope Church on the 26th."

Our Army retreated from Dalton on the night of the 12th and the morning of the 13th of May, and, as just cited, Cantry's Division of three thousand (3000) was at Resaca on the 9th, and Loring's of five thousand (5000) on the 11th.

Thus, we discover fourteen thousand two hundred (14,200) infantry, and thirty-nine hundred (3900) cavalry under General Jackson, moving en route to Dalton, prior to the 9th of May; and that the head of Polk's column, which was Canty's Division, joined General Johnston's left, at Resaca, on that date. which facts seemingly indicate that there were at least some troops "within easy direction" of this General on the 6th of May. Let us, however, for the present, adhere to the question of the strength and losses of his Army.

Since the cavalry increased so greatly in number after the 1st May, it is reasonable to suppose that the infantry and artillery likewise augmented after that date, and before we left Dalton.

It was telegraphed over the country that General Sherman was about to advance. This information induced quite a large number of absentees to return; as General Hardee and myself had noticed this increase a short time before the retreat began, the subject was mentioned between us when discussing the approaching campaign, and we found by comparison that we had as many as forty-two thousand five hundred (42,500) effective infantry and artillery in our two corps, exclusive of not less than six hundred (600) effectives in the reserve artillery. If a return had been made up on the 10th or 12th of May, the number I have stated would have appeared upon it.

The following extract from a letter of Colonel W. H. Sellers,

Assistant Adjutant General of Hood's Corps, Army of Tennessee, dated October 20th, 1872, Galveston, Texas, furnishes evidence of the correctness of my statement in this regard:

"I cannot be as positive regarding the strength of your command during the operations at and from Dalton to Atlanta as I could wish. My recollection, however, is that you mustered twenty-one to twenty-two thousand (21,000 to 22,000) effectives at Dalton and Resaca, at which latter point some diminution occurred in casualties, and in desertions on the night of our retreat on Cassville.

Hardee's Corps was the largest in the Army, and numbered about two thousand (2000) more than my corps.

As previously stated, the assertions of General Wigfall as to Johnston's strength and losses may safely be regarded as correct; and General Johnston furnishes evidence of his satisfaction therewith by inserting in his Narrative the speech delivered by Wigfall in the Senate Chamber of the Confederate States. This Senator, in his estimate of the strength of Polk's Corps,* says "it amounted to less than nineteen thousand (19,000) men."

Colonel Douglas West, of New Orleans, La., who was at that time Assistant Adjutant General of Polk's Corps, says on November 13th, 1869, in answer to a letter from me in regard to the strength of that corps when it joined General Johnston, "We bore on the rolls an aggregate of about twenty-four thousand (24,000) present." General Johnston acknowledges to have received eighteen thousand one hundred (18,100) from that source.

We now have forty-two thousand five hundred (42,500) in Hood's and Hardee's Corps at Dalton, exclusive of six hundred reserve artillery; and about nineteen thousand (19,000) in Polk's Corps, which was marching rapidly to that point, together with eight thousand four hundred and ten (8410) of Wheeler's Cavalry, exclusive of Jackson's.

*Johnston's Narrative, page 591.

We find, by this summary, seventy thousand five hundred (70,500) effectives—a number in excess of that which is stated in my official report. This number of troops, however, did not, at that time, embrace all the available forces which were subject to the order of General Johnston.

The following extract from Major General G. W. Smith's official report attests the presence of over three thousand (3000) Georgia State troops, which could have been, in May, 1864, assembled at Dalton, in the event the Commanding General of our Army had desired to offer battle, when in possession of Rocky-faced Ridge:

"HEADQUARTERS, GEORGIA MILITIA,
"MACON, GA., *September 15th, 1864.*

"GENERAL J. B. HOOD,
"Commanding Army of Tennessee, near Lovejoy Station:

"GENERAL:—My appointment was dated 1st June. I took command a few days thereafter, relieving General Wayne, who returned to the duties of his office as Adjutant and Inspector General of the State. The force then in the field was composed entirely of State officers, civil and military. They had been formed into two brigades of three regiments each, and one battalion of artillery, making in all a little over three thousand (3000). The command had reported for duty to General J. E. Johnston, and had been ordered to guard the crossings of the Chattahoochee river from Boswell Bridge to West Point."

These troops were far superior to those usually found in the ranks of the militia, as they were composed of the civil and military officers of the State, and were possessed of more pride and intelligence. They could have performed noble service in well-constructed redoubts in Mill Creek and Snake Creek Gaps; would have proved the equal of regulars in those positions, and have allowed General Johnston the grand opportunity to attack Sherman with his main Army, by passing over the northern slope of Rocky-faced Ridge.

We find, as above stated, forty-two thousand five hundred (42,500) in Hardee's and Hood's Corps; nineteen thousand (19,000) in Polk's Corps; eight thousand four hundred and ten (8410) in Wheeler's immediate command, and three thousand (3000) Georgia State troops under General Wayne; thus forming a grand total of seventy-three thousand five hundred and ten (73,510) effectives "at and near Dalton," either marching to or in readiness, to be promptly massed at that point.

It will be observed that I have estimated the total effective cavalry at ten thousand one hundred and thirty-nine (10,139), whereas, Major Falconer, in his return of the 10th of June (page 574, Johnston's Narrative), acknowledges ten thousand five hundred and sixteen (10,516); also, that I have made no allowance for the return to duty of some of the wounded, prior to the passage of the Etowah, nor for the killed and wounded of the cavalry, prisoners, stragglers, and deserters in the two estimates at New Hope Church and Adairsville.

I have been compelled to make these various estimates in order to demonstrate the actual strength of General Johnston's Army, since he furnished the War Office with no returns after the 1st of May until June 10th, and since he, as stated by his own Adjutant General, took with him all the books and records of the Army when he relinquished the command. Major Falconer is sustained in his statement in regard to the removal of the records to Macon, by the following declaration, which prefaces the diary of Brigadier General Shoupe:

"*Memoranda of daily movements and events in the Army of Tennessee, kept by Brigadier General* F. A. SHOUPE, *assigned to duty as Chief of Staff by orders from General* HOOD, dated July 24th, 1864:

"No records were turned over by former chief of staff; therefore, the records of the office embrace only the administration of General Shoupe."

Major Falconer, in referring to General Johnston's last return of the 10th of July, says: "The report was made under General Johnston, and signed by General Hood. On the 18th of July the command was turned over to General Hood."

He estimates the force turned over to me on the 18th of July, eight days after this return, at fifty thousand six hundred and twenty-seven (50,627) effectives, assuming that no losses occurred from the 10th to the 18th of July. The last eight days General Johnston commanded the Army. This supposition is not reasonable, since eight thousand and twenty-one (8021) were lost the thirty days previous to the 10th. Owing to the change of commanders under such extraordinary circumstances, surely from two to three thousand deserted during the interval. Therefore, I estimated the number of the Army of Tennessee turned over to me on the 18th of July at forty-eight thousand seven hundred and fifty (48,750), which estimate I arrived at through my chief of staff, Brigadier General Shoupe, who was with me at the time I made my official report. I also placed his losses at twenty-two thousand seven hundred and fifty (22,750), and his strength at seventy thousand (70,000) effectives, when I knew them to have been in excess thereof. My desire, however, was not to over-estimate either.

My attention having been called to the exaggerated statements of Federal officers in regard to my losses around Atlanta, it will be seen that I telegraphed the War Department on the 18th of August that General Johnston turned over to me forty-nine thousand and twelve (49,012) effectives. This must have been the assumed estimate of Major Falconer at the time, as no return was made up on the 18th of July.

Having established the strength of the Army to have been over seventy thousand (70,000) effectives after General Polk's Corps joined, it only remains to be shown that these reinforcements were "available."

General Johnston asserts in his Narrative, page 304, "On the 5th the Confederate troops were formed to receive the enemy." On the next page, referring to the same date (May 5th), he states, "In the evening a telegram from Lieutenant General Polk informed me that he had been ordered to join the Army of Tennessee with all his infantry"; also, "At daybreak on the 7th the Federal Army moved forward; * * * in the afternoon the Federal Army placed itself in the front of the Confederate line, its right a little south of Mill Creek Gap, and its left near the Cleveland road."

General Wigfall furnishes the following information obtained from the War Office: *

> "It was not till the 4th of May that General Polk was ordered to move with Loring's Division and other available force at your command to Rome, Georgia, and thence unite with General Johnston." On the same page he states that on the 6th of May the following dispatch was sent to General Cooper, at Richmond, by General Polk from Demopolis, Alabama, "My troops are concentrating and moving as directed."

It will be seen that on the 4th of May, Polk's Army had been ordered to join the Army of Tennessee; was concentrating and moving forward rapidly by rail from Demopolis on the 6th, having but a short distance to march; and that General Sherman did not take up his position in front of Rocky-faced Ridge until the afternoon of the 7th of May.

Between the two Armies arose, I might say, a high wall of stone, as the name Rocky-faced Ridge indicates. The Confederate position was one of the strongest to be desired; it was necessary to hold but two gaps in the mountains: Mill Creek and Snake Creek. The approach of the Federal Army down the railroad from Chattanooga, in lieu of down the road from Cleveland, rendered the position the more secure, inasmuch as General Johnston would not have had a stone wall between him and his adversary, had General Sherman advanced by the latter route, where the country is open towards Cleveland. I have always thought General Sherman did not wish to accept a pitched battle, or he would have moved upon Dalton from that direction. His advance by the Chattanooga road, and, subsequently, in front of Rocky Face, convinces me that his intention was to initiate the policy of wasting our strength, which he so effectually carried out in the campaign

* Johnston's Narrative, page 590.

from Dalton to Atlanta. He came to the position of all others most favorable, if our commander could have been induced to hold his ground, viz: in front of Rocky-faced Ridge.

General Polk was not far distant from Dalton, when it is considered that eight thousand (8000) of his troops were at Resaca on the 9th and 11th, and that he in person was in Dalton on the 12th. General Johnston could well have awaited the arrival of the whole of this Army, since it required so small a force to hold Mill Creek and Snake Creek Gaps, as previously stated, and practically demonstrated by General Sherman's use of them, after these mountain defiles fell into his possession.

When en route to Tennessee, during the campaign in the Fall of 1864, the Confederate Army, after having captured the troops stationed at Dalton, attempted to march through Mill Creek Gap; it was prevented from so doing by a squad of men posted within a little fort, covered with railroad iron, and which had been constructed of logs of large size, around which was thrown up an embankment of earth to protect the troops against field artillery; port holes had been cut so as to allow the men to fire in all directions, and especially upon the line of the railroad. It was reported to me that field artillery had little or no effect upon this impromptu fortification, and that when the men charged up to it they could not find an entrance; therefore, it could not be taken without much loss of time, and considerable cost. Major Kinloch Falconer was severely wounded while experimenting with this little fortress, which occasioned the Army to march several miles around it. At a later hour, however, after the order to move forward had been issued, this block-house was surrendered to a detachment of our troops.

It is not often the case that all the troops to be brought into action are assembled beforehand at the precise point where a great battle is imminent. On the contrary, when two armies are approaching each other, each commander manœuvres, and, generally, one is forced to keep his adversary in check day, issued his farewell order that memorable night. In despite of my repeated and urgent appeals to him to pocket all despatches from Richmond, to leave me in command of my own corps, and to fight the battle for Atlanta, he deserted me the ensuing afternoon. He deserted me in violation of his promise to remain and afford me the advantage of his counsel, whilst I shouldered all responsibility of the contest.

I reiterate that it is difficult to imagine a commander placed at the head of an Army under more embarrassing circumstances than those against which I was left to contend on the evening of the 18th of July, 1864. I was, comparatively, a stranger to the Army of Tennessee. Moreover General Johnston's mode of warfare formed so strong a contrast to the tactics and strategy which were practiced in Virginia, where far more satisfactory results were obtained than in the West, that I have become a still more ardent advocate of the Lee and Jackson school. The troops of the Army of Tennessee had for such length of time been subjected to the ruinous policy pursued from Dalton to Atlanta that they were unfitted for united action in pitched battle. They had, in other words, been so long habituated to security behind breastworks that they had become wedded to the "timid defensive" policy, and naturally regarded with distrust a commander likely to initiate offensive operations.

The senior Corps Commander considered he had been supplanted through my promotion, and thereupon determined to resign, in consequence, I have no doubt, of my application to President Davis to postpone the order transferring to me the command of the Army; he however, altered his decision, and concluded to remain with his corps.

The evening of the 18th of July found General Johnston comfortably quartered at Macon, whilst McPherson's and Schofield's Corps were tearing up the Georgia Railroad, between Stone Mountain and Decatur; Thomas's Army was hastening preparations to cross Peach Tree creek, within about six miles of Atlanta; and I was busily engaged in hunting up the positions of, and establishing communication with Stewart's and Hardee's Corps, since I did not know where they were posted, when General Johnston disappeared so unexpectedly and left me in this critical position.

Not till I read Sherman's Memoirs, was I aware of McPherson's so close proximity to Atlanta at an early hour on the 18th of July. In truth, a few enterprising scouts thrown out that afternoon from his columns, in the direction of the Macon Railroad might have captured my predecessor on his retreat to Macon.

Sherman says (vol. II, pages 71, 72):

> "On the 18th all the Armies moved on a general right wheel, Thomas to Buckhead, forming line of battle facing Peach Tree creek; Schofield was on his left, and McPherson well on towards the railroad between Stone Mountain and Decatur, which he reached at 2 p. m. of that day, about four miles from Stone Mountain, and seven miles east of Decatur, and there he turned toward Atlanta, breaking up the railroad as he progressed, his advance guard reaching Decatur about night, where he came into communication with Schofield's troops, which had also reached Decatur."

It thus appears that on the afternoon of the 18th the enemy was in Decatur, almost at the gates of Atlanta. This intelligence must have been communicated to General Johnston by the cavalry, after he left me to ride into the city with the promise to return toward evening, as he was virtually Commander-in-Chief up to the moment of his sudden departure. I had consumed a great portion of that day in vain endeavors to adjust the difficulties in the way of his retention in command, by earnest representations to him, on one hand, and, on the other, by telegraphing to Richmond in the hope of accomplishing this object. Although he had published his farewell order the night previous, I had not, owing to the foregoing reasons, assumed command. He had agreed to issue orders in my name, and, in reality, I did not become Commander-in-Chief until about night of that day, when I received information of his departure.

the 6th, the date mentioned in my official report, since our Army was still at Dalton on the 12th, when nearly one-half of Polk's Corps had already joined Johnston's left at Resaca on the 9th and 11th of May. Therefore, the trivial point raised by these two gentlemen is of little or no consequence.

CHAPTER V.

REPLY TO GENERAL JOHNSTON—TRANSFER FROM THE VIRGINIA TO THE WESTERN ARMY—DALTON, RESACA, ADAIRSVILLE, AND CASSVILLE.

BEFORE I relate the embarrassing circumstances under which I assumed command of the Army of Tennessee, I shall state certain facts connected with my transfer from the Virginia to the Western Army early in February, 1864, and reply to statements of General Johnston, in reference to operations near Resaca, Cassville, and New Hope Church.

The War Department had been anxious that an offensive campaign into Tennessee and Kentucky be initiated in the early Spring of 1864, and made a proposition to General Johnston to reinforce him with Polk's troops, then in Mississippi, and Longstreet's Corps, in East Tennessee. Johnston, at the appointed time, was expected to move forward and form a junction with these troops. The President and General Bragg, and also General Lee, were desirous that the offensive be assumed, and an attempt be made to drive the Federals to the Ohio river, before a large Army could be concentrated to move against us. The following* extract from a letter of General Bragg to General Johnston, dated March 12th, 1864,

* Johnston's Narrative, page 292.

will show the number of men proffered the latter, if he would carry out the expressed wishes of the authorities at Richmond:

"It is needless, General, for me to impress upon you the great importance, not to say necessity, of reclaiming the provision country of Tennessee and Kentucky; and from my knowledge of the country and people, I believe that other great advantages may accrue especially in obtaining men to fill your ranks.

"The following forces, it is believed, will be available, if nothing should occur to divert them, viz.:

	Infantry.	Artillery.	Cavalry.	TOTAL.
Your own command...........	33,000	3,000	5,000	41,000
General Martin's cavalry now en-route to you................	....	....	3,000	3,000
From Lieutenant General Polk..	5,000	....	....	5,000
From General Beauregard......	10,000	....	....	10,000
From General Longstreet's command......................	12,000	2,000	2,000	16,000
	60,000	5,000	10,000	75,000

"It is proposed to hold the reinforcements ready, and to put them in motion just as soon as you may be able to use them. To throw them to the front now, would only impede the accumulation of supplies necessary for your march."

I here give the subjoined extract from a letter of General Bragg, addressed to me at the close of the war:

"NEAR LOWNDESBORO', ALABAMA,
"*17th December, 1865.*

"MY DEAR GENERAL:—In addition to the Army of Tennessee, then at Dalton, the General commanding there was offered, for an offensive campaign, Polk's Corps from Mississippi and Alabama, Longstreet's Corps from East Tennessee, and a sufficient number from Beauregard's command in South Carolina and Georgia, to make up seventy-five thousand (75,000) effective infantry. The cavalry with these commands would have numbered at least ten thousand (10,000), and the artillery six thousand (6000)—Total, ninety-one thousand (91,000). Besides the effective, *so reported*, there were not less than fifteen thousand (15,000) able-bodied men bearing arms, but reported on *extra duty*, such as clerks, cooks, mechanics, laborers, teamsters, etc.,—one-half of whom, at least, could at any time be placed in battle without impairing the efficiency of the Army. * * To furnish the means, all other Armies were, for the time being, to be subordinated to the Army of Tennessee. Even General Lee, with the Army of Virginia, was to give up Longstreet's Corps, and remain on the defensive.

"Yours truly,

"BRAXTON BRAGG."

The President had thus agreed to afford General Johnston every facility in his power for the execution of the proposed plan of operations; and it was with the understanding we were to enter upon an active campaign that I consented to leave the Army of Northern Virginia, with which I had served since the outbreak of the war.

On the evening of my arrival at Dalton, on or about the 4th of February, I repaired to General Johnston's headquarters, and reported to him for duty. During our interview, in his room alone, he informed me that General Thomas was moving forward, and he thought it might be best for us to fall back and take up some position in rear of Dalton. I at once told him that I knew nothing of the situation or of the object of General Thomas's move from Ringgold, but that we could, at least, hold our position a sufficient length of time to compel the enemy to develop his plan. The Federals, in a few days, fell back to Ringgold, having merely made a feint, in order to cover some movement then being made in Mississippi.

This was my introduction to the Army of Tennessee; albeit not calculated to inspire or encourage military ardor,—since it was proposed to retreat even before the enemy became in earnest—I nevertheless laid before General Johnston the plan to join Polk's Army and Longstreet's Corps on the march into Tennessee, gave him assurance that the authorities in Richmond would afford him every assistance, and informed him, moreover, that General Lee favored the projected campaign.

General Johnston immediately took the ground that he did not very well know the country through which it was proposed to pass to the rear of the enemy; that there were difficulties to be encountered, etc., etc.; he desired Polk's and

Longstreet's forces to join him at Dalton, where, this large Army being concentrated, he considered he should be left to decide and act for the best; in other words, be left to move forward, stand his ground or retreat, as might seem most expedient.

To this demand, General Lee was unwilling to accede; he was reluctant to give up Longstreet's Corps, unless for the purpose of active work and dealing hard blows, in the performance of which task it had already so often distinguished itself. The War Department objected to the withdrawal of Polk's Army from Mississippi, until active operations were to commence, as by such a movement one of the best regions of country for supplies would be abandoned to the enemy. Thus matters stood until the 7th of March, when, still anxious for the offensive, I wrote to President Davis, suggesting that Polk join us at Dalton, and we move forward to make a junction with Longstreet.

I will here incidentally remark that the following is the only correspondence I remember ever to have had with the authorities at Richmond, while occupying a subordinate position, and its object was the furtherance of General Johnston's wishes:

"DALTON, GEORGIA, *March 7th, 1864*.

"*To His Excellency, President* JEFFERSON DAVIS.

"I have delayed writing to you so as to allow myself time to see the condition of this Army. On my arrival, I found the enemy threatening our position. I was, however, delighted to find our troops anxious for battle. He, the enemy, withdrew after taking a look, and is now resting with his advance at Ringgold.

"I am exceedingly anxious, as I expressed to you before leaving Richmond, to have this Army strengthened, so as to enable us to move to the rear of the enemy and with a certainty of success. An addition of ten or fifteen thousand (10,000 or 15,000) men will allow us to advance. We can do so anyhow by uniting with Longstreet.

"But so much depends upon the success of our arms on this line, that I thoroughly appreciate the importance of collecting together all the forces we possibly can, in order to destroy the Army under General Grant. We should march to the front as soon as possible, so as not to allow the enemy to concentrate, and advance upon us. The addition of a few horses for our artillery will place this Army in fine condition. It is well clothed, well fed, the transportation is excellent and in the greatest possible quantity required.

"I feel that a move from this position, in sufficient force, will relieve our entire country. The troops under Generals Polk and Loring having united with the forces here, and a junction being made with General Longstreet, will give us an Army of sixty or seventy thousand (60,000 or 70,000) men, which I think should be sufficient to defeat and destroy all the Federals on this side of the Ohio river.

"I sincerely hope and trust that this opportunity may be given to drive the enemy beyond the limits of the Confederacy. I never before felt that we had it so thoroughly within our power. He, the enemy, is at present weak, and we are strong. His Armies are far within our country, and the roads open to his rear, where we have a vast quantity of supplies.

"Our position in Virginia can be securely held by our brave troops under General Lee, which will allow us to march in force from our centre, the vital point of every nation.

"You find, Mr. President, that I speak with my whole heart, as I do upon all things in which I am so deeply interested. God knows I have the interest of my country at heart, and I feel in speaking to you that I am so doing to one who thoroughly appreciates and understands my feeling.

"I am eager for us to take the initiative, but fear we will not be able to do so unless our Army is increased.

"Believe me, with great respect, your friend and obedient servant.

"J. B. HOOD."

The same difficulty here arose as before mentioned: unwillingness upon the part of the authorities at Richmond to order Polk from Mississippi, and reluctance on the part of General Lee to give up Longstreet, before it was positively ascertained that active operations were to commence. As to the time of such active operations, General Johnston would not specify. So stood this important matter in abeyance, until the 13th of April when I addressed General Bragg the following letter:

[Private.]

"Dalton, Georgia, *April 13th, 1864.*

"My Dear General.

"I received your letter, and am sorry to inform you that I have done all in my power to induce General Johnston to accept the proposition you made to move forward. He will not consent, as he desires the troops to be sent here, and it be left to him as to what use should be made of them. I regret this exceedingly, as my heart was fixed upon going to the front, and regaining Tennessee and Kentucky. I have also had a long talk with General Hardee. Whilst he finds many difficulties in the way of our advancing, he is at the same time ready and willing to do anything that is thought best for our general good. He has written a long letter to the President, which will explain his views.

"When we are to be in better condition to drive the enemy from our country, I am not able to comprehend. To regain Tennessee would be of more value to us than half a dozen victories in Virginia.

"I received a letter from General R. E. Lee yesterday, and he says, 'you can assist me by giving me more troops or driving the enemy in your front to the Ohio river. If the latter is to be done, it should be executed at once.' * * *

"Since McPherson's Corps has moved up from the lower Mississippi to join the Army of the Potomac or that of the Cumberland, would it not be well for General Polk's troops to unite with this Army, as we should then be in a condition to reinforce General Lee, in case it should be necessary?

"Yours truly,

"J. B. HOOD.

"To General Braxton Bragg."

It will be seen that I was still urgent for an offensive campaign, and even counselled that Polk be ordered to Dalton, in the hope that we would finally advance, and join Longstreet in Tennessee. At the same time, I was not unmindful of the great danger of leaving Mississippi open to the enemy, before being able, by unmistakable preparations for a forward move, to attract the undivided attention of the Federals in the West. Unfortunately, however, assumed difficulties set forth by General Johnston prevented the execution of, in my opinion, one of the most important campaigns projected during the war; and one fact is certain, whatever may be said contrariwise, President Davis offered every possible inducement towards its execution, and had, in regard to the wisdom of the proposed operations, the support of General Robt. E. Lee.

I cannot name one of Lee's Lieutenant Generals who would not have met this proposition from the War Department with that spirit of co-operation which is so essential in time of war. Moreover, any officer possessed of even a part of that heroic self-reliance so characteristic of Lee and Jackson, would not only have gladly accepted the ninety-one thousand (91,000) men, but, having secured a competent Quarter Master, would soon have found the necessary transportation; would have sent a dispatch to Richmond that he was moving forward, and, God willing, would take from the enemy all else needed to equip the army. Such might have been the result, instead of unremitting demands, upon the part of General Johnston, for an outfit equal to that of United States troops, visions of insuperable difficulties, and vacillations unending.

I am now convinced that even the concentration of Polk's Army and Longstreet's Corps, at Dalton, would in no manner have altered the ensuing campaign. If I had had a conception of the operations from Dalton to Atlanta, naught but the most peremptory orders could have induced me to have left General Lee.

General Johnston, in reference to the operations around Resaca, makes the following remarks: *

"Major General Stevenson had early in the day, and with Lieutenant General Hood's approval, assumed the position from which he had been recalled the night before. Here he was directed by the Lieutenant General to place a field battery in a position some eighty yards in front of his line of infantry. Before the necessary arrangements begun for its protection were completed, he was directed by General Hood to open its fire. This was no sooner done, than so impetuous an attack was made upon it that the guns could not be drawn back to the main line of the division. After a very sharp contest, the enemy was driven beyond the battery by the well directed fire of Brown's and Reynolds's brigades, but found shelter in a ravine not far from it. From this position their musketry commanded the position of the battery equally as well

*Johnston's Narrative, page 313.

as that of the Confederate infantry, so that neither could remove the guns, and they were left between the two armies until night."

He asserts,* " no material was lost by us in this campaign but the four field pieces exposed and abandoned at Resaca by General Hood."

I was anxious to occupy a commanding position in my front before the enemy obtained possession thereof. Stevenson's Division, of my corps, as well as the Federals, were moving rapidly towards this point. A battery was placed in position in order to check the enemy, and allow my troops time to reach the ground, the object of contention. Whilst these four guns accomplished the desired aim, the concentrated fire of a number of Federal batteries forced the gunners to withdraw, and leave them between the lines of the two armies, which were very close together at that point. They were finally abandoned on the night of our retreat from Resaca, simply from the fact that I found upon consultation with Colonel Beckham, my chief of artillery, and Major General Stevenson, one of my division commanders, that I had more guns than were required for the number of men in my command; and, as the order to retreat had been given, it was deemed better to yield them to the enemy than to sacrifice one or two hundred men in reclaiming them. I think my action, in this instance, will meet not only the approval of the military, but also of the civilized world.

The whole matter was laid before General Johnston, and the guns were abandoned with his concurrence; at least such is my recollection. Moreover, I am informed by Captain Sweat that these guns belonged to his command, and that they were four old iron pieces, not worth the sacrifice of the life of even one man.

The following letter from General Johnston's chief of ordnance, Colonel Oladowski, is at variance with the statement that " no material was lost by us in the campaign but the four field pieces, exposed and abandoned at Resaca by General Hood:"

---

*Johnston's Narrative, page 351.

"MOBILE, *29th May, 1874*.

"GENERAL B. BRAGG.

"GENERAL:—I answered your telegram day before yesterday; hasten to-day to answer your letter, received this morning. I read attentively General Johnston's Narrative, and it seems to me he tried to vindicate himself at the cost of others. His statement of losses is based upon report of his Medical Director. I wonder how a doctor could know about deserters, stragglers, prisoners, etc. I am extremely disappointed.

"I cannot positively state the reduction of his Army from Dalton to Atlanta, but I believe it was about nineteen thousand (19,000) muskets. * * * * As to the deficiency of ammunition, it is a romance. I left full supplies on hand at the time General Hood took command. * * *

"Very respectfully, your obedient servant,

"H. OLADOWSKI."

The above is in answer to a letter written at my request by General Bragg. It is impossible that we should have lost twenty-five thousand (25,000) men from Dalton to Atlanta, and, at the same time, no material save four field pieces. After the muskets of the killed and wounded were gathered and turned into the Ordnance Department, nineteen thousand (19,000) is about the proportion that might be expected to have been lost through stragglers, deserters, and prisoners, during such a campaign.

Colonel Oladowski will be remembered by many soldiers of the Army of Tennessee, not only as a gentleman of high character, but also as an officer who was faithful and exact in the performance of his duty.

I regret to find it necessary to notice this small affair, upon which General Johnston seems to lay so great stress, whereas he might, in his Narrative, have furnished the future historian with important matter, had he given an account of the miraculous escape of his Army at Resaca, when, under cover of darkness, we marched over bridges commanded by the enemy's guns, and were thus extricated from the pocket, or, I may say, *cul de sac*, in which he had placed us, with two deep and ugly streams, the Connasauga and Oostenaula, in our immediate rear.

Of this historical fact there is no mention whatever in General Johnston's book; and I shall always believe the attack of Stevenson's and Stewart's Divisions, therein described (page 311), together with our return to our original position on the following day, saved us from utter destruction by creating the impression upon the Federals that the contest was to be renewed the next morning. They were thus lulled into quiet during that eventful night of our deliverance. It was upon this occasion General Polk remarked to an officer of high rank, now residing in New Orleans, that our escape seemed almost a miracle.

In regard to operations around Cassville, General Johnston states:*

"Next morning (19th of May), when Brigadier General Jackson's report showed that the head of the Federal column following the railroad was near Kingston, Lieutenant General Hood was directed to move with his corps to a country road about a mile to the east of that from Adairsville, and parallel to it, and to march northward on that road, right in front. Polk's Corps, as then formed, was to advance to meet and engage the enemy approaching from Adairsville, and it was expected that Hood's would be in position to fall upon the left flank of these troops as soon as Polk attacked them in front. An order was read to each regiment, announcing that we were about to give battle to the enemy.

"When General Hood's column had moved two or three miles, that officer received a report from a member of his staff, to the effect that the enemy was approaching on the Canton road, and in rear of the right of the position from which he had just marched. Instead of transmitting his report to me, and moving on in obedience to his orders, he fell back to that road and formed his corps across it, facing to our right and rear, towards Canton, without informing me of this strange departure from the instructions he had received.

"I heard of this erratic movement after it had caused such a loss of time as to make the attack intended impracticable; for its success depended on accuracy in timing it. The intention was therefore abandoned."

This is, indeed, a grave charge on which I am arraigned, and, if sustainable, I should have been deprived of my com-

*Johnston's Narrative, page 321.

mand at the time. An officer of high grade who would prove so incompetent as to fail to initiate a battle ordered and planned by his Commander-in-Chief, is worthy of the severest censure. I will, however, give a simple statement of the facts.

The three Corps Commanders, especially General Polk and myself, urged General Johnston, soon after our arrival at Cassville, to turn back and attack Sherman at Adairsville, as we had information of a portion of his Army having been sent to cross the Etowah, in order to threaten our communications south of that river. The opportunity was the more favorable, because of an open country and good roads, which would have enabled the Army to move rapidly and force the Federals, whilst divided in their forces, to accept a pitched battle, with rivers in their rear. This he declined to do, as stated in my official report; in no part of his Narrative, however, can be found the slightest allusion to this matter.

On the following day, Howard's Corps having been reported on the Ironton road (the country road referred to), I asked his authorization to march my command across an open field, and attack this detachment of the enemy, in case the report was correct. He consented.

I received no orders for battle as related by General Johnston, nor were the Corps Commanders brought together and given explicit instructions, verbal or written, as is usual and necessary upon the eve of a general engagement, although he had published, soon after our arrival at Cassville, a general order to the effect that he intended to fight. I was merely granted the privilege of doing what I had requested; the assertion, therefore, of General Johnston, that I had been ordered to move to the country road and be in readiness to attack in flank when Polk engaged the enemy in front, is as erroneous as it is inexplicable.

In accordance with his authorization, I put my troops in motion. After riding some distance in advance, I found the country road in possession of our own dismounted cavalry, and turned to meet the head of my column, then pushing forward, when, suddenly, in an open space adjoining the Canton road, appeared a line of the enemy advancing upon our flank, and in rear of the position we had just left. This line opened both an artillery and musketry fire upon my troops, who were marching by flank across the same open range. General Hindman was ordered to throw out from his division a line of skirmishers to develop the enemy approaching from this so unexpected direction, and suffered a loss of several men killed and wounded.

About this juncture General Mackall, Chief of General Johnston's staff, rode up in great haste, and said in a most excited manner that General Johnston desired I should not separate myself too far from General Polk. I called his attention to the enemy, in sight, advancing in the open field, and told him I had been in person to the Ironton road; had found it in possession of our cavalry, and could, therefore, at any moment, easily form on the right of Polk. His reply was "Very well," or words to that effect.

Polk had not moved from the position in which I had left him, as I had his right in full view; and, surely, if General Johnston had intended that I should have been in position to attack in flank when Polk engaged the enemy in front, Polk would already have been moving forward, or I would have been ordered by Mackall to remain on the country road till he (Polk) advanced and engaged the enemy. I was within two hundred yards of the Ironton, or country road, when General Mackall overtook me.

In an appended note,* General Johnston affirms, in regard to the appearance of the enemy on the Canton road, "the report upon which General Hood acted was manifestly untrue."

If General Johnston be right, I am not only to blame for not fighting in accordance to instructions (which were never given), but also for allowing myself to be deceived by an imaginary line of the enemy advancing from an unexpected direction, in an open field, and firing upon my troops about

* Johnston's Narrative, page 321.

four or five hundred yards distant. On the other hand, if my report be true, it was my duty, upon the appearance of the enemy in line of battle, accompanied by artillery, and moving upon our flank and rear, not only to report the same to my commander, but to halt, and force the enemy to develop his strength and object, even if I had been given orders by General Johnston to deliver battle, which orders, I reiterate, were never issued to me, for, be it remembered, I had merely been authorized to carry out my own suggestion.

The following letters from Major J. E. Austin, one of the most gallant and efficient officers of the Army of Tennessee, and from the Honorable Taylor Beattie, of the State of Louisiana, both gentlemen of honor and prominent position, show whether or not the report I made to General Johnston was "manifestly untrue."

"NEW ORLEANS, *May 26th, 1874.*

"MY DEAR GENERAL:—In the disposition of the Army under General Joseph E. Johnston, at Cassville, Georgia, as he states, for attack, I commanded the extreme right of the skirmish line in front of your corps. In your movement to the north, across the open field, on that day, I covered your front and right, and my command was left to observe the enemy when a part of your corps was thrown across the Canton road. The enemy were in force in my front, with artillery and infantry, and developing toward our right. About an hour and a half before nightfall the enemy broke through the skirmish line of an Alabama brigade posted to my left, and moved rapidly in rear of my main line, which was threatened in front. My reserve was, consequently, disposed to meet this movement in the rear, and encountered and repulsed the enemy in a short and severe engagement on and near the Canton road; but gathering reinforcements, he moved further to the rear, until I was completely isolated and cut off from your corps.

"To extricate my command, I had to move to the right, fighting all the while in front and rear, until darkness put an end to hostilities. By making a detour of eight or nine miles in the night over a country devoid of road, I was enabled to rejoin your corps, massed in column about two o'clock in the night, and just in time for my wearied men to participate in the retreat across the Etowah, which was shortly after begun.

"From my observations, I am forced to believe that General Johnston makes an error in his book in discrediting the presence of the enemy on your right, while you were moving to the north, across the large open field, to get in position. If my memory serves me, your extreme right flank was not covered by cavalry at all, as is not only usual but most essential in a movement such as you were making, and you must have had to rely for information of the enemy in that quarter from your staff and escort.

"I am, General, very truly yours,

"J. E. AUSTIN,

"Major commanding Austin's Battalion Sharp Shooters."

"PARISH OF ASSUMPTION, *March 29th, 1874.*

"GENERAL J. B. HOOD.

"DEAR SIR: I remember very well the occurrences at Cass Station, or Cassville, during the campaign of 1864. During that campaign I kept a diary, which I have just examined to refresh my memory. At the risk of being somewhat tedious, I will state all I know of that affair. Your corps being in the rear of the Army, entered Cassville about 12 m., on the 18th of May, 1864. Yourself and staff (on which I was active as volunteer aid) came in last and found the Army massed by brigades in front of Cassville,—that is, between that town and the approaching enemy. So we remained all night. Next morning I first heard of the celebrated battle order of General Johnston. I refer to that order in which it was announced that our retreat was ended, and that, if the enemy continued his advance, battle would then and there be given him. I thought it strange, if such were the determination, that the Army had not been placed in line of battle the evening before, when they would, at least, have been more comfortable. After our breakfast we rode to General Johnston's headquarters, where you remained some half hour or more. We then rode to General Hindman's Division, which was immediately placed in motion to take up its position in line of battle, as I supposed.

"As we marched across an open valley towards the range of hills which I understood to be the line to be occupied by our forces, you being in front with General Hindman and I just behind, one of the soldiers called my attention to a dark line off to our right, saying they were Yankees. I called your attention to the fact, but you said it could not be so, but must be our cavalry, if it was a body of men. Falling back, the same soldier (to whom I had said I thought the dark line was a fence or hedge), said that they were throwing out skirmishers in our direction, and I at once called your attention to that fact. You halted the division, and ordered General Hindman to send out a body of skirmishers to find out who they were. In a few minutes a sharp skirmish was in progress, and several of our men were wounded and killed in your immediate proximity. I recollect very distinctly that five men were hit at one time by the fragments of a shell, which exploded not more than twenty-five yards from where you were sitting on horseback.

"About this time General Mackall, chief of staff to General Johnston, rode up, apparently much excited, and spoke with you. Of course, I cannot say what took place, but soon after,—indeed, at once, and in his presence,—our direction was changed, and we proceeded to take up a line of battle on the range of hills immediately in rear of Cassville.

"There we remained all day, the enemy erecting batteries in front and in flank of us, and enfilading our line. This fire of the enemy's artillery was harassing in the extreme, as it seemed to come from all directions except our immediate rear, and we made little or no reply. Late that evening you took us (your staff) from the left of your corps to the extreme right (which was also the right of the Army) and back. I shall long remember that ride as one of the most disagreeable it has ever been my fortune to take, being, as we were, continually under a heavy cross-fire of artillery. As soon as the ride was over, you proceeded to headquarters of the Army, and on your return you notified us that we would retreat at midnight across the Etowah, and gave us the necessary orders.

"It was afterwards said in the Army, indeed, I am not certain that a report by General Johnston was not published to that effect, that the retreat was ordered because you and General Polk had declined to fight —or rather had given an opinion adverse to battle. I can say, after four years' experience of war, that I am satisfied that no soldiers in the world could have held the line then occupied by your corps during the next day, unless the enemy had been very remiss in taking advantage of his position.

"I have heard that General Johnston, in his history of the war, says you were mistaken about the enemy being immediately on your flank on May 19th, 1864. He was misinformed by whomsoever gave him this idea, for, as I have said, several of our men were killed and wounded by this raking fire in your immediate proximity, and before our Army was in line of battle.

"Very truly yours,

"TAYLOR BEATTIE,

"Late Colonel C. S. A."

The foregoing statements prove the report characterized by General Johnston as "manifestly untrue," to be manifestly true. Five thousand witnesses, moreover, could be produced to testify to the truth of my assertion. It would, indeed, seem strange that, after battling three long years with the same enemy, wearing the same uniform, and bearing the same colors, I should be so grossly deceived as to make a false report, especially when I had a full view of this same enemy in an open field, within a distance of four to five hundred yards.*

I did not fall back, and form across the Canton road, as General Johnston states; his chief of staff overtook me too soon to allow this movement; in accord with General Mackall's instructions, I marched back to join Polk's right, which had remained in the same position I had left it. Whilst Major Austin was still engaged with this same enemy on the Canton road, and my corps was nearing the line occupied by General Polk on the ridge in front of Cassville, orders were issued for the Army to fall back to the ridge in rear of the town.

This position was commanded by the ridge we were about to abandon, and a greater portion thereof was exposed to an enfilade fire of the enemy's artillery. General F. A. Shoupe, General Johnston's chief of artillery, advised the Confederate commander of this fact before the Army was ordered to occupy that line, as I stated in my official report. General Johnston does not, if I remember correctly, refer to General Shoupe's monition in his own official report, which he has failed to publish in his Narrative, although the latter purports to be his contribution for the use of the future historian; this report should be one of the most important records of his military career, for the omission of which, however, he apologizes by stating that it had been published by the Confederate Government.

In reference to the position established on the ridge in rear of the town, this General writes:†

"The Federal artillery commenced firing upon Hood's and Polk's troops soon after they were formed, and continued to cannonade until

---

* Since the foregoing was penned, General Carson, of the Federal Army, who is now engaged in writing an account of General Hooker's operations, informs me that it was a portion of General Butterfield's command which appeared on the Canton road, and fired into my column.

† Johnston's Narrative, page 323.

night. Brigadier General Shoupe, chief of artillery, had pointed out to me what he thought a weak point near General Polk's right, a space of a hundred and fifty or two hundred yards, which, in his opinion, might be enfiladed by artillery placed on a hill more than a mile off, beyond the front of our right—so far, it seemed to me, as to make the danger trifling. Still, he was requested to instruct the officer commanding there to guard against such a chance by the construction of traverses, and to impress upon him that no attack of infantry could be combined with a fire of distant artillery, and that his infantry might safely occupy some ravine in rear of this position during any such fire of artillery."

It will be seen that the artillery of the enemy opened upon Polk's and my troops soon after they were formed; that according to the above statement, General Shoupe pointed out only a small portion of our line which might be enfiladed by artillery; also that General Shoupe was requested to instruct the officer there commanding to guard against this evil by the construction of traverses. The truth is, General Shoupe reported to General Johnston that a large portion of the ridge he proposed to occupy in rear of Cassville, would be enfiladed by the Federal artillery; in other words, that the position of the line subsequently occupied by Polk and myself would be so enfiladed. This is in substance what I stated in my official report; and this statement was written by General Shoupe himself, at present the Reverend Dr. Shoupe, of Sewanee, Tennessee. The subjoined letter is confirmatory of my assertion:

"SEWANEE, TENNESSEE, *June 3d, 1874.*

"DEAR GENERAL:—With regard to the point you mention, I have a very distinct recollection. I pointed out the fact to General Johnston that his line would be enfiladed before the troops were posted, and suggested a change of position to obviate the trouble.

"The General replied that the troops could not hope to be always sheltered from fire, and that they must make the best of it by traversing.

"As soon as the enemy got into position, my fears were fully verified. The line, at that point, fell back from the crest of the ridge, but was poorly sheltered even upon the slope. I should say that there was as much as a quarter of a mile badly exposed to the enemy's fire.

"General Polk was present at the time the conversation between

General Johnston and myself took place, and strongly supported my objections.

"I am indeed sorry to have my name mixed up in the difference between General Johnston and yourself, but I do not see that I can decline to reply to such questions as you please to ask.

"With high regard, I am, &c.,

"F. A. SHOUPE."

The memory of General Johnston must assuredly have become very treacherous to have forgotten, not only the remonstrances of General Shoupe, but the earnest opposition of General Polk; and, moreover, to have reduced the distance of a quarter of a mile to one hundred and fifty or two hundred yards. (See Map.)

General Johnston must have placed but little reliance upon traverses in this position, since, if I am not mistaken, he made no mention of them in his official report.

At all events, traverses could not, in this instance, have proved a sufficient protection to the troops.

After our lines had been enfiladed for one or two hours before sunset, as General Shoupe had pre-admonished General Johnston, Polk and I decided, upon consultation, to see the Commanding General and apprise him of our real condition; to state also that, whilst our position was as good as we could desire to move forward from and engage the enemy in pitched battle, the line we held was unsuited for defence; and if he did not intend to assume the offensive the next morning, we would advise him to change his position. This is the sum and substance of our suggestion, or recommendation, to the Commander of that Army, viz.: that if he did not intend to fight a pitched battle, we would advise him to change our position for one better suited for defence.

This suggestion would seem to be in unison with the spirit of our urgent recommendation only the day previous to turn upon Sherman, and give him general battle at Adairsville, and but poorly harmonizes with the following:*

---

* Johnston's Narrative, pages 323, 324.

"On reaching my tent soon after dark, I found an invitation to meet the Lieutenant Generals at General Polk's quarters. General Hood was with him, but not General Hardee. The two officers, General Hood taking the lead, expressed the opinion very positively that neither of their corps would be able to hold its position next day; because, they said, a part of each was enfiladed by Federal artillery. The part of General Polk's Corps referred to was that of which I had conversed with Brigadier General Shoupe. On that account they urged me to abandon the ground immediately, and cross the Etowah."

I have already stated that the corps commanders, especially Polk and myself, urged Johnston only the day previous to march back and attack Sherman at Adairsville; that his own chief of artillery reported the position Polk and I occupied as unsuited for a defence, before our retreat from the ridge in front of the town; I have *proved* that the enemy appeared on the Canton road, according to my report—which report General Johnston declared to be "manifestly untrue." Now since Polk and I most earnestly urged Johnston, only the day previous, to move forward and attack Sherman, does it not seem strange that we should be insisting on retreat the following night? Admitting, however, for the sake of argument, that Johnston intended to fight when in position on the untenable ridge in rear of Cassville, this intention could only have been based upon the vain hope that Sherman would march across the valley, and through the town to attack his entrenchments. The Federals would never have made an assault from this direction, as the country toward Canton was open, and favorable to an attack upon our right flank. Humanity itself should have prompted this way of approach, in order to spare the women and children of the town. Again, even in the event Polk and I had consented to subject our troops to a heavy enfilading fire of artillery, may I not ask—especially as a part of Sherman's Army, I think Schofield's Corps, was then reported to be moving across the Etowah to threaten our communications south of this stream, and a similar movement had dislodged us already from Dalton and Resaca, and in fact dislodged us from every position between Dalton and Atlanta—

how long is it supposed we would have remained at Cassville? I leave the answer to every fair minded man.

This is the history of the much talked of affair at Cassville, in connection with which it is affirmed that Johnston wished to fight, but Polk and I were not inclined to do so.*

General Johnston, as evidence that without any proviso we advised him to retreat, quotes a statement to that effect of General Hardee and his chief of staff, neither of whom were present during our long discussion. If I did remark after the interview had closed, and Johnston had decided to cross the Etowah, that, if attacked, Polk would not be able to hold his line three-quarters of an hour, nor I mine two hours, the remark could have had but a distant bearing upon the question. A single observation could not have unfolded all which occurred during this prolonged meeting.

If General Polk were to rise up from his grave, he would be astounded at the suppression of the most important part of the testimony in relation to these facts, by which, during a period of ten years, the impression has been sustained that we are both to blame for not permitting General Johnston to fight, when he was so desirous to deliver battle. With the foregoing statement, I do at this day and hour, in the name of

---

* The following letter from Dr. A. M. Polk, son of General Polk, at that time aide-de-camp to his father, sustains the truthfulness of this representation of facts:

"NEW YORK, *June 17th, 1874.*

"DEAR GENERAL:—I have just read your correction of General Johnston's statements in regard to my father's connection with the 'Cassville affair.'

"Pray accept our sincere thanks, not only for the correction, but also for the manner in which it is expressed. He was killed so soon after, he left no written statement of the matter; but from conversations I held with him I know his position to have been just as you state it: not willing to stand there and wait for the enemy to attack us, but more than willing to take the initiative in bringing on a general engagement.

"With much respect,

"I am most truly yours,

"A. M. POLK."

truth, honor and justice, in the name of the departed soul of the Christian and noble Polk, and in the presence of my Creator, most solemnly deny that General Polk or I recommended General Johnston, at Cassville, to retreat when he intended to give battle; and affirm that the recommendation made by us to change his position, was throughout the discussion coupled with the proviso: *If he did not intend to force a pitched battle.*

CHAPTER VI.

REPLY TO GENERAL JOHNSTON—CASSVILLE.

WHEN the preceding chapter was written, setting forth my most positive denial of General Johnston's statements in regard to that which he avers to have been said by General Polk and myself, at Polk's headquarters, during this important council; and when I charged General Johnston with the suppression of the most important part of the recommendations made to him by each of us, I was under the impression that only Johnston, Polk and I were present in the room during the discussion. Fortunately, however, the complete vindication of my assertion has arisen from a source I little expected. In addition to the strong evidence adduced by the letters of General Shoupe and Doctor Polk, I am favored with the subjoined full and explanatory letter from a gentleman of no less position than that of chief engineer of a corps d'armee, and who was present, in the room, during the council of war held by Johnston, Polk, and myself, with map and measurement of angles of the position in question:

"NEW YORK, *June 25th, 1874.*

"DR. W. M. POLK, *288 Fifth Avenue, New York.*

"DEAR SIR:—In reply to your note of the 20th inst., asking me to give you my recollection of the circumstances in regard to the retreat of the

Confederate Armies from Cassville, Georgia, to the south side of the Etowah river, I will state the facts as connected with myself, as follows:

"At the time when the Confederate Armies of Tennessee and Mississippi, under the command of General J. E. Johnston, and the Federal Army under General Sherman, were manœuvring in the neighborhood of Cassville, I had nearly completed my journey from Demopolis, Alabama, to that town to join Lieutenant General Polk, commanding the Army of Mississippi, who was with General Johnston in that vicinity. I had crossed the country in company with a part of that command. I arrived at Cassville railway station about half-past three or four o'clock in the afternoon of the 19th of May, 1864, and met Colonel Gale, of our staff, who informed me that the Lieutenant General desired to see me as soon as I arrived. I passed on without delay to his headquarters, about half a mile east of the railway station, and met General Polk at the door of the cabin used for headquarter purposes. I entered immediately, and he placed a skeleton map before me, giving the surrounding country, and pointed out the positions of the Confederate forces, and the known and supposed locations of the Federals, giving such additional information as to enable me to fully understand the actual condition of affairs. This was done rapidly. He then requested me to go at once and examine the extreme right of his line, as he considered it untenable for defence.

"1st. He desired me to form an opinion if, by constructing a rifle pit, his line could be held against such an attack as might be reasonably expected in the morning.

"2d. To carefully examine that part of the line enfiladed, to see if it was possible to construct traverses to enable him to hold the position on the defensive.

"3d. To examine the ground immediately in his front in reference to advancing, and to note in reference the positions then occupied by the Federal batteries in front and to the right of Lieutenant General Hood's line.

"4th. If those batteries to the front and right of Hood's line could be taken by a special movement. These explanations, noting them down, and getting a tracing of the skeleton map, required about thirty minutes, and I started for that part of the line in question; General Polk impressing upon me the necessity of reaching that part of the line as soon as possible, as I would only have about two hours of daylight to make the examinations. Furnishing me with a fresh horse, one of his own, and the necessary guides from his escort, I reached the ground in fifteen minutes. I was instructed to return as soon after dark as possible, for, if necessary, an invitation would be sent to General Johnston to come to his (Lieutenant General Polk's) headquarters. Lieutenant General Hood, I think, was with General Polk when I left. Arriving upon the line of battle, I found Major General French's Division, Army of Mississippi, located on the extreme right of that Army, and occupying the part of the line in question. To his right was the line of Lieutenant General Hood's Corps, Army of Tennessee, forming the extreme right of the Confederate infantry forces. The crest of the ridge occupied by French's Division was about one hundred and forty feet above the plain, or valley, in which the town of Cassville is located. This ridge is cut across by a ravine of about fifty feet deep, its sides rising from its bottom, on either side, at about 30 degrees. The location of this ravine on French's line was five or six hundred feet to the left of his extreme right. To the left of this ravine, for twelve or fifteen hundred feet, the crest of the ridge was entirely open, as was to the rear for eight hundred or one thousand feet. There were a few scattered trees of stunted growth in and about the ravine. The remaining portion of General French's line to the left and to the rear was timbered, as also to the front for seven or eight hundred feet, increasing in depth towards the left. The ground to the front of the left half of his line descended about one hundred and forty feet for half a mile, continuing on to Cassville about one and a quarter miles to the northwest of his left. The ground in front of the right half of his line descended about a hundred feet on the left, and eighty feet on the right for a distance of half a mile on the left, and a quarter of a mile on the extreme right. Then ascending to eighty feet on the left, and a hundred on the right to a ridge opposite, and due north.

"This opposing ridge passed on a line about 23 degrees south of west, forming an angle with General Polk's line of defence of about 25 degrees, and forming something less of an angle with Lieutenant General Hood's line. This opposite ridge was occupied by the enemy, their left resting on a point about a mile and a quarter northeast on a prolongation of General Polk's line, and from half a mile to three-quarters of a mile in front of Lieutenant General Hood's, and passing on to the westward at a distance of about half a mile to one and a quarter miles north of General Polk, and in front of his extreme right. The line occupied by the enemy on the opposite ridge was from twenty to forty feet higher than the position of General Hood's line, and from forty to sixty feet higher than General Polk's. The batteries of the enemy were posted on the most prominent and available points along their ridge, extending for a mile from their extreme left towards their right, reaching a point to the north and front of General Polk's extreme right,

Map 1

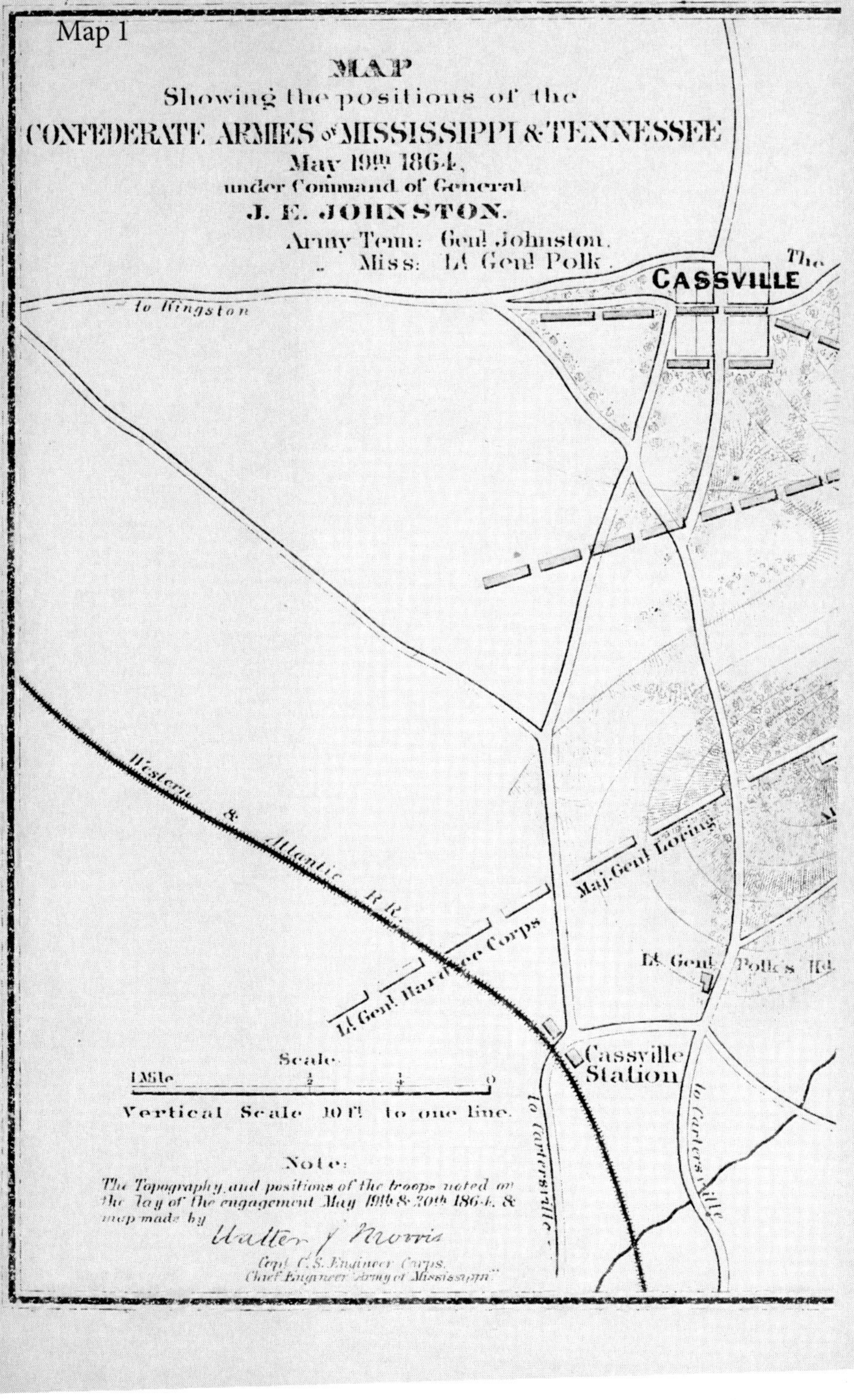
MAP
Showing the positions of the
CONFEDERATE ARMIES of MISSISSIPPI & TENNESSEE
May 19th 1864,
under Command of General
J. E. JOHNSTON.
Army Tenn: Genl Johnston.
" Miss: Lt Genl Polk.
CASSVILLE
The
to Kingston
Western & Atlantic R.R.
Maj. Genl Loring
Lt Genl Hardee Corps
Lt Genl Polk's Hd
Cassville Station
to Cartersville
to Cartersville
Scale
1 Mile ½ ¼ 0
Vertical Scale 10 ft to one line.
Note:
The Topography and positions of the troops noted on the day of the engagement May 19th & 20th 1864, & map made by
Walter J Morris
Capt. C. S. Engineer Corps.
Chief Engineer "Army of Mississippi"

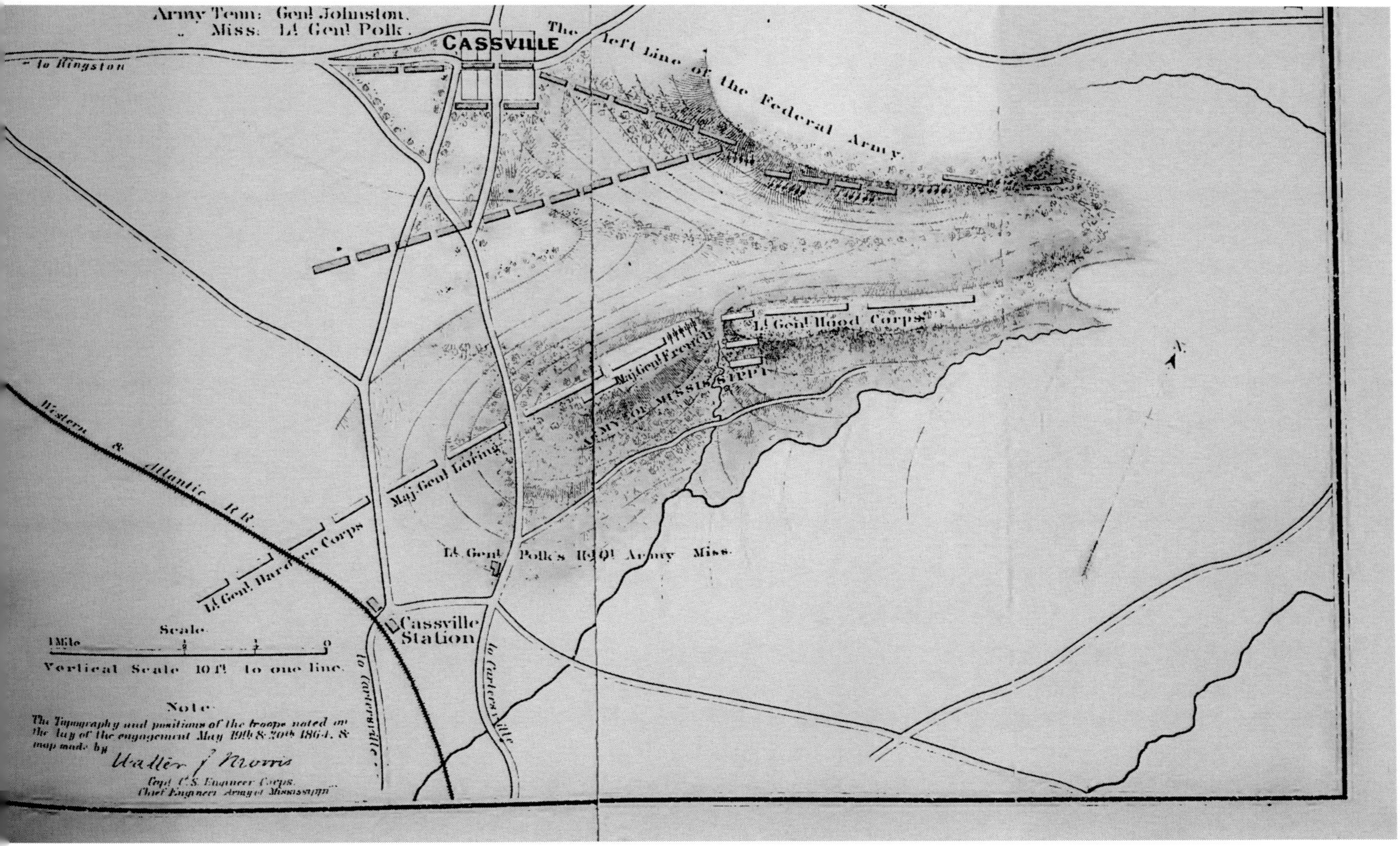

Army Tenn: Gen! Johnston.
" Miss: Lt Gen! Polk.
CASSVILLE
to Kingston
The left line of the Federal Army.
Lt Gen! Hood Corps
Maj. Gen! French
ARMY OF MISSISSIPPI
Maj. Gen! Loring
Lt Gen! Hardee Corps
Western & Atlantic R.R.
Lt Gen! Polk's H.Qt! Army Miss.
Cassville Station
to Cartersville
to Cartersville
Scale
1 Mile
Vertical Scale 10 ft to one line.
Note
The Topography and positions of the troops noted on the day of the engagement May 19th & 20th 1864. & map made by
Walter J Morris
Capt C.S. Engineer Corps.
Chief Engineer Army of Mississippi

and directly in front of the raviné and open part of French's line. The batteries enfiladed and cross fired upon the entire open crest from 45 degrees to 60 degrees, and with a plunging fire of from twenty to sixty feet and sweeping through the ravine, and across the rear of the ridge to a distance of about a thousand feet. This rear fire being still more plunging than that on the crest.

"There was no cover for the men within a reasonable distance to the crest, for from the extreme positions of the left batteries of the enemy, it would not be necessary for them to cease firing during the attack until their infantry had reached a line very close to the crest of the ridge occupied by General Polk's command.

"The extreme left, or eastern batteries of the enemy, necessarily enfiladed a considerable portion of General Hood's line.

"Having made these examinations and noted them down, I formed the following opinions:

"1st. That the right of the line occupied by Lieutenant General Polk's command could not be held, as it then was, nor could it be held by constructing a rifle pit along the crest.

"2d. That traverses would be of no avail either for the rifle pits upon the crest or as a covered way to the rear, as such traverses would cover nearly the entire surface.

"3d. That it was extremely hazardous for Lieutenant General Polk to advance his line to make an attack upon the enemy while their batteries held the positions they then occupied.

"4th. As to forming any opinion as to the taking of these left batteries of the enemy by a special flank movement, this I could not do, as I was unable to examine to the right of Lieutenant General Hood's line, as it had grown dark. But judging from the stream, as located on the skeleton map, there must have been a very narrow ridge to approach the enemy upon their left.

"At the time I arrived about the centre of General Polk's right where the open crest of the ridge commenced, I found a very heavy enfilading and cross fire going on from the enemy's batteries. There were but a few sentinels remaining upon the crest, the main body of men, intended to occupy this part of the line, were compelled to withdraw to the right and left at the foot of the ridge, out of sight, but not out of range of the enemy's batteries.

"I found that Major General French had one or two batteries in position upon the part the line near the ravine, and while they were coming into their positions, and before the guns could be unlimbered, from one to two horses from each piece were killed. On my return over this part of the line, about dark, the fire from the enemy had nearly ceased.

"Having completed the reconnoissance, I returned to Lieutenant General Polk's headquarters, just after dark.

"I placed before him my sketches and notes, and explained to him substantially these facts. General Polk sent at once to ask General Johnston to come to his headquarters. Lieutenant General Hood was already with General Polk. General Johnston arrived about 9 o'clock. I remained in the cabin during the conversation as to holding the position then occupied or advancing or retiring the Armies to the south of the Etowah river, about seven or eight miles to our rear.

"Lieutenant General Polk expressed himself convinced that he could not hold his line against attack, and that Major General French, who occupied that part of his line in question, was of the same opinion as was his (General Polk's) engineer officer (myself), who had examined the position and reported that traverses would be of no avail. Lieutenant General Hood stated that he was also convinced that neither he nor General Polk could hold their lines for an hour against such an attack as they might certainly expect in the morning—these Generals both advocating to the Commanding General to take the offensive and advance on the enemy from these lines. In reference to this proposed forward movement, General Johnston's attention was particularly called to the advantages of taking possession of the positions occupied by the batteries of the enemy on their extreme left, either by a special flank movement or by prompt action at the time when the Confederate lines would be advanced. Lieutenant General Polk expressed himself entirely willing and ready to co-operate with General Hood to accomplish this object. After some moments of silence, General Johnston decided to withdraw the Armies to the south of the Etowah. Soon after this, Lieutenant General Hardee arrived. General Johnston informed him of this decision to cross the river, stating that Generals Polk and Hood had informed him that they could not hold their lines. Lieutenant General Hood then re-stated the reasons, and said that General Polk could not hold his line an hour. Nor could he, Hood, hold his two hours if attacked in the morning. Lieutenant General Polk again explained the facts as existed in reference to his line, and stated his willingness to assume the offensive at any time, then or in the morning, rather than to await the attack of the enemy in his (Polk's) present position. Upon these points Lieutenant Generals Polk and Hood entirely agreed, urging the offensive rather than await the enemy.

"Lieutenant General Hardee made but few, if any, remarks that I heard. After a few moments General Johnston gave the orders for the armies to move to the south side of the Etowah. Lieutenant General

Polk called to his A. A. General to issue orders to his Division Commanders. This was about 10.30 or 11 o'clock.

"The orders to Major General Loring, Army of Mississippi, were given me to deliver; also one to him to order to report to me an officer with three hundred (300) men to occupy the exposed part of Major General French's line, as soon as his command was withdrawn.

"I was instructed by General Polk to place this detail along that part of the line, and keep up such fires as would indicate the presence of the withdrawn command, and to cut timber and drive stakes to indicate that works were being thrown up, and to remain there until daylight and observe the movements of the enemy before leaving. I went at once to General Loring's headquarters on the left of the Cassville road, saw that General, and delivered the orders; obtained the officer and detail, and arrived at General French's line about half-past eleven o'clock, and found that command ready to move; by twelve o'clock (midnight), they had withdrawn and the detail was posted with a few men out in front. It was a calm, clear starlight night, and the position of the enemy upon the opposite ridge was clearly seen, without their fires which could be traced along their line, and the cutting of timber could be distinctly heard and located. In addition to the enemy's location upon the crest of the ridge, and passing there or just in front of the town of Cassville and on to the southwest, there were also strong indications of an advance line upon the plane nearer to the foot of the ridge occupied by us, and their chopping and driving rails was very distinct, and their voices occasionally could be heard.

"The work of the detail was kept up through the night. At daylight I instructed the officer to assemble his men to the rear. During this time of preparing to leave the line, I closely observed the enemy and his positions through a very strong field glass. I found that many of their batteries along the ridge had been advanced, and their principal and somewhat entrenched line appeared to leave the ridge at a point about a mile east of Cassville, and passing to the southwest fully a half-a-mile in front of their lines of the previous afternoon. It appeared that the enemy had been aware of the movement of the Confederate Armies, and their line advanced during the night, was now vacated and there were trains and artillery moving to the west upon the Kingston road, and solid bodies of infantry were moving in the same direction.

"The detail having been assembled, I placed them upon a by-road to Cassville Station on the main road to Cartersville. I instructed the officer to proceed to the south side of the Etowah river by way of the Cartersville bridge, and to report back to his Division Commander. I passed on to cross the river at the same point, arriving there about half-past ten o'clock, and found the Army of the Mississippi nearly over to the south side, which was completed by noon.

"Very truly yours,

"WALTER J. MORRIS,

"Late Captain Engineer Corps, C. S. A

"Chief Engineer, Army of Mississippi

"N. B.—Enclosed herewith you will find a map made by me from my notes taken at the time of reconnoissance.*

"Yours, etc.,

"W. J. M."

---

* Map of Cassville, page 113.

CHAPTER VII.

REPLY TO GENERAL JOHNSTON—NEW HOPE CHURCH—KENNESAW MOUNTAIN—RETREAT ACROSS THE CHATTAHOOCHEE—JOHNSTON RELIEVED FROM COMMAND.

GENERAL JOHNSTON, touching the operations of his Army near New Hope Church says:*

"We found, next morning, that the Federal line extended much further to our right than it had done the day before. Polk's Corps was transferred to the right of Hood's. * * * The Federal troops extended their entrenched lines so rapidly to their left, that it was found necessary in the morning of the 27th to transfer Cleburne's Division of Hardee's Corps to our right, where it was formed on the prolongation of Polk's line. Kelly's Cavalry, composed of Allen's and Hannon's Alabama brigades, together less than a thousand (1000) men, occupied the interval, of half-a-mile, between Cleburne's right and Little Pumpkin-vine creek. * * * * Between 5 and 6 o'clock in the afternoon, Kelly's skirmishers were driven in by a body of Federal cavalry, whose advance was supported by the Fourth Corps. * * * * As soon as the noise of this contest revealed to Major General Cleburne the manœuvre to turn his right, he brought the right brigade of his second line, Granberry's, to Kelly's support, by forming it on the right of his first line. * * * The Fourth Corps came on in deep order, and assailed the Texans with great vigor, receiving their close and accurate fire with the fortitude always exhibited by General Sherman's troops in the actions of this campaign. * * * The contest of the main body

*Johnston's Narrative, pages 328, 329, 330.

of the Fourth Corps with Granberry's brigade was a very fierce one. * * * They (the enemy) left hundreds of corpses within twenty paces of the Confederate lines."

It is strange the author of this Narrative should offer the above, and, in fact, nearly all he has written on pages 328–29–30–31 as a contribution to the historian, when he commits the unpardonable error of placing Polk's Corps during the whole of this "affair near New Hope Church" in the identical position occupied by my corps.

I was not only on the right, where he places Polk, but sent to him for a good division, with the message that Howard's Corps was moving rapidly to turn my right flank, which was the right of the infantry of our Army; that I had extended my lines as far as possible. He sent Cleburne's Division to report to me. General Cleburne was given by me most explicit instructions in regard to the formation of his forces on the right of my corps. He was directed to place his troops in a column of brigades, in the rear of my immediate right, which was the right of Hindman's Division, with Granberry's brigade in rear of the column, so as to bring it on our extreme right when deployed into line; he was also instructed to allow the Federal cavalry to reconnoitre and find our right. Similar orders were given to our own cavalry. As Howard's Corps advanced, Cleburne was directed to deploy quickly into line; the Federals thus came in contact with a solid line of infantry, in lieu of finding the open space on our flank, which existed at the time of the reconnoissance of the Federal cavalry.

I shall ever remember the enthusiasm and transport of the gallant Cleburne at the time of this though small engagement, yet most brilliant affair of the whole campaign.

The proof of the correctness of my statement respecting the above operations will be found in the following extract from a short report, written at my dictation by a young officer of my staff, and which, as it conflicts with General Johnston's own Narrative, is unaccountably inserted by him on pages 585 and 586:

"On the morning of the 26th, the enemy found to be extending their left. Hindman's Division was withdrawn from my left, and placed in position on my right, the enemy continuing to extend his left. Major General Cleburne, with his division, was ordered to report to me, and was massed on Hindman's right. On the morning of the 27th, the enemy known to be extending rapidly to the left, attempting to turn my right as they extended. Cleburne was deployed to meet them, and, at half-past 5. p. m., a very stubborn attack was made on his division, extending to the right, where Major General Wheeler, with his cavalry, dismounted, was engaging them. The assault was continued with great determination upon both Cleburne and Wheeler until after night, but every attempt to break their lines was gallantly repulsed. About 10 o'clock at night, Brigadier General Granberry, with his brigade of Texans made a dashing charge on the enemy, driving them from the field, their killed and wounded being left in our hands. During this engagement, two or three hundred prisoners were captured, all belonging to Howard's Corps."

At the end of this hastily written field report I add, "I enclose Major General Cleburne's report, and will forward others as soon as received." Every soldier of and above the rank of captain knows that no officer sends forward his reports of battle, save through his commanding officer at the time of the engagement. Therefore General Cleburne brought his report of this "affair" to me, who commanded him at the time, in lieu of forwarding it through Lieutenant General Hardee to whose corps he was attached.

Again, in reference to operations near New Hope Church, the author of this remarkable Narrative writes as follows, page 333:

"When the three Lieutenant Generals were together in my quarters that day (the 28th), as usual, Lieutenant General Hood suggested that we should make an attack upon the Federal Army, to commence on its left flank. The suggestion was accepted, and the three officers were desired to be ready for battle next morning. Lieutenant General Hood was instructed to draw his corps out of the line to the rear, and to march during the night around our right, and form it facing the enemy's left flank, somewhat obliquely to his line, and to assail that flank at dawn next day. Polk and Hardee were instructed to join in the battle successively, obliquely to the present formation, when the progress made on the right of each should enable him to do so.

"We waited next morning for the signal agreed upon—the musketry of Hood's Corps—from the appointed time until about 10 a. m., when a message from the Lieutenant General was delivered to me by one of his aides-de-camp, to the effect that he had found Johnston's Division, on the Federal left, thrown back almost at right angles to the general line, and entrenching; that, under such circumstances, he had thought it inexpedient to attack, and asked for instructions. I supposed, from the terms of this message, that Hood's Corps was in the presence of the enemy, and that, his movement and position being known to them, they would be prepared to repel his assault as soon as he could make it, after his aide-de-camp's return. If the attack had been expedient when Lieutenant General Hood's message was dispatched, the resulting delay, by enabling the enemy to reinforce the threatened point and complete the entrenchments began, made it no longer so. He was therefore recalled."

Before I withdrew from the right of the Army which rested on Little Pumpkin-vine creek, with Cleburne's Division still on my extreme right and under my orders—*i. e.*, before I withdrew on the night of the 28th of May from the position General Johnston erroneously assigns General Polk during the 26th, 27th and 28th, I received information from General Wheeler's cavalry stationed on Cleburne's right, just across Little Pumpkin-vine creek, that the enemy had its left flank beyond this stream, in a position which was exposed by reason of the difficulty of passage back to the main body of their Army; and that if I could withdraw that night, the 28th, and get in position by early morning, I might attack this corps or division thus exposed, and destroy it before it could recross Little Pumpkin-vine creek or receive reinforcements. This information reached me on the morning of the 28th, after Cleburne's repulse of the enemy on the afternoon and night of the 27th, as before mentioned.

Encouraged by this favorable opportunity of dealing the enemy a hard blow, I instantly repaired to General Johnston's headquarters and asked his permission to withdraw my corps at dark from our extreme right, and attack this exposed flank next morning. He answered that it might result in a general engagement; to which I replied that, if I were able to destroy one portion of the enemy before it could be reinforced, it would give us greatly the advantage if a general battle ensued; that Hardee and Polk could be in readiness to come to my assistance, if necessary.

Having obtained his consent, couriers were dispatched for the two remaining corps commanders, Hardee and Polk, who shortly joined us. They were instructed to hold their corps in readiness for action the next day, as I was going to march that night, upon the above report from Wheeler's cavalry, and attack the left flank of the enemy,—provided I found it as reported; in other words, the whole of the proposed movement was to depend upon the enemy's left flank remaining as represented.

Polk was then, for the first time, ordered to my position—the right of the Army—and, accordingly, I withdrew after night and took up my line of march with guides from Wheeler's cavalry. Just about dawn, as we were approaching the place where the enemy was reported to be in an exposed position, I received from the same cavalry a message to the effect that I need proceed no further, as the Federals had during the night, drawn back their left flank, recrossed Little Pumpkin-vine creek, and were entrenched. From a feeling of insecurity, they had recrossed to the side of the creek I had left the evening previous, thereby placing between the opposing forces a swamp and difficult stream to cross, in addition to entrenchments on the opposite bank. An attack upon the enemy after he had recrossed to the side of the creek I had left the night before, would have been extreme rashness, especially, since I had had an opportunity during one or two days previous to my move from the position I occupied at the time Cleburne was on my right, to make a similar assault without having to encounter the obstacles of a swamp and a creek. Our cavalry had evidently seen the folly of attacking the Federals across this creek, and, therefore, advised me to proceed no further. I reported these facts to General Johnston, and was ordered to return.

The following extract from a letter dated May 22d, 1874, received from General Wheeler, General Johnston's Chief of Cavalry, will show that the enemy was heavily entrenching the night of my march around our right flank:

> * * * * "I recall the movement to attack the enemy's left flank with your corps and my cavalry, which, I think, was on the night of the 28th. I remember you sending for me on the morning of the 29th, and telling me why you did not attack, which was owing to a change in position of the enemy and their invariable custom of entrenchment. I remember that the enemy were cutting down trees during the night, which was one of their favorite plans of strengthening and even building works, especially in so densely wooded a country. I cannot recall what officer was in charge of the scouts or in command of the brigade immediately in front of the enemy's left flank."

I have a strong impression that the officer to whom General Wheeler refers was the gallant General Kelly, who was afterwards killed in battle.

It might be supposed, upon reading General Johnston's recital of this his second attempt to fight, that I was ordered to assault the enemy under any circumstances, and that I was again the cause of battle not having been delivered. Never within my history have I been ordered to fight and have failed to obey instructions. I have never experienced pleasure in being shot at, but I have always endeavored to do my whole duty; and, although I have been charged with recklessness in regard to the lives of my men, I had sufficient caution to know that some positions should not be attacked, such as the one occupied by the enemy after recrossing Little Pumpkinvine creek. However, had General Johnston given me orders to attack at all hazard, I would have done so. It is true I went into battle under protest at Gettysburg, because I desired to turn Round Top Mountain; but, notwithstanding, I was true in every sense of the word to the orders of my commander till, wounded, I was borne from the field.

During three years' service, under Generals Lee, Jackson, and Longstreet, I was never charged with being too late in any of the many battles in which I was engaged, before reporting for duty with the Army of the West.

When General Johnston said "as usual," I suggested that we attack the left flank of the enemy. I presume he had in remembrance Lieutenant General Polk's and my urgent recommendation that he turn upon and attack Sherman at Adairsville, just before he placed his Army upon the untenable ridge in rear of Cassville, with women and children of the town between the two armies, and of which recommendation he is so careful to make no mention.

When I retrace these facts and circumstances, I cannot think General Johnston in earnest when he states that he intended, or desired to fight at the different points mentioned; moreover, it must seem strange to my comrades of the Virginia Army that I, who had always been ready and willing to do my duty, should have undergone so complete a change under General Johnston, during the last year of the war. In truth, I had nowise altered in my nature; and I will add that no General ever received more thorough co-operation of his corps commanders than did General Johnston during his campaign from Dalton to Atlanta. He was on cordial terms with each of us, and it should be borne in mind that the animus displayed towards General Polk and myself, never became apparent till after I was assigned to the command of the Army of Tennessee, and the noble Polk had been laid in his grave nigh two months. General Johnston was then residing in Macon, Georgia, where he wrote his official report, in which were brought forward, for the first time, these unjust and false accusations.

If I was so little to be relied upon, and had given cause for complaint successively at Resaca and Cassville, why did he entrust to me the important operations at New Hope Church, from which it was supposed a general engagement might ensue. The truth is, he possessed no real cause of complaint, and, I reiterate, he had the full co-operation of his Lieutenants. No matter what were the views held by them touching his mode of handling an army, they were all sufficiently good soldiers to forego, in the presence of even one of their own staff officers, any remark which might tend to destroy confidence in their leader.

I will cite a historical fact illustrative of this spirit of discretion and forbearance, which will be peculiarly interesting as it has never, to my knowledge, been made public.

Just before leaving New Hope Church, his three corps commanders were assembled alone, at night, in his quarters—then a little cabin near the church—when General Johnston suggested Macon as being the place to fall back upon. If I remember rightly, this suggestion was received in silence, for I cannot recall the reply of one of us at the moment. I well remember, however, after we had left the presence of General Johnston, and were riding through the darkness of the night to our respective headquarters, that the unanimous sentiment expressed on this occasion was to this effect: "In the name of Heaven, what is to become of us? Here we are with the depots for recruits drained, from Mobile to Richmond, all the troops having been sent either to us or to General Lee, in Virginia; our Army fifty or sixty miles from Dalton, no general battle fought, and our Commander talking of Macon, one hundred miles beyond Atlanta, as being the place to fall back upon!"

This gloomy outlook brought about the comparison touching our losses up to that period, and to which I have previously referred. We finally separated; each rode off to his own tent; and, howsoever, dispirited, I am confident not one of us so far lost sight of that co-operation so essential in time of war, as to speak one word which would convey a suspicion of General Johnston's contemplated retreat to Macon.

Shortly after this occurrence, the Army occupied the line at Kennesaw Mountain, the last stronghold of the many sharp ridges passed over during our retreat. It was to the left of this point, on Pine Mountain, that we lost the brave and magnanimous Polk, and with him much of the history of this remarkable campaign.

The Confederate Army had remained on the defensive about thirty days at Kennesaw Mountain, when Sherman resorted to a ruse he had learned from experience would prove effective: he sent a few troops to make a rumbling sound in our rear, and we folded up our tents, as usual, under strict orders to make no noise, and, under cover of darkness, marched to and across the Chattahoochee, upon the flat plains of Georgia.

After our passage of this river, on the night of the 9th of July, Sherman moved rapidly to the eastward and across the Chattahoochee, some distance above Peach Tree creek. He formed a line parallel to this creek, with his right on the river, and approached Atlanta from the north, whilst Schofield and McPherson, on the left, marched rapidly in the direction of Decatur to destroy the railroad to Augusta.

General Johnston thus relates the sequel:*

"On the 17th, Major General Wheeler reported that the whole Federal Army had crossed the Chattahoochee. * * * The following telegram was received from General Cooper, dated July 17th: 'Lieutenant General J. B. Hood has been commissioned to the temporary rank of General, under the late law of Congress. I am directed by the Secretary of War to inform you that, as you have failed to arrest the advance of the enemy to the vicinity of Atlanta, far in the interior of Georgia, and express no confidence that you can defeat or repel him, you are hereby relieved from the command of the Army and Department of Tennessee, which you will immediately turn over to General Hood.' * * * General Hood came to my quarters early in the morning of the 18th, and remained there during the day. Intelligence soon came from Major General Wheeler, that the Federal Army was marching toward Atlanta, and, at General Hood's earnest request, I continued to give orders through Brigadier General Mackall, Chief of Staff, until sunset."

---

* Johnston's Narrative, pages 348, 349, 350.

About 11 o'clock, on the night of the 17th, I received a telegram from the War Office, directing me to assume command of the Army. This totally unexpected order so astounded me, and overwhelmed me with sense of the responsibility thereto attached, that I remained in deep thought throughout the night. Before daybreak I started for General Johnston's headquarters, a short distance from which I met Lieutenant General A. P. Stewart, one of my division commanders, who had been recommended by me, and recently promoted to the rank of corps commander to replace General Polk.

We rode on together to General Johnston's quarters, which we reached shortly after dawn. I at once sought the Commanding General, and inquired into the cause of this order. He replied he did not know; the President had seen fit to relieve him. I then insisted he should pocket that dispatch, leave me in command of my corps, and fight the battle for Atlanta; at the same time I directed his attention to the approach of General Sherman, and alleged that the enemy, unless checked, would in a few days capture the city.

To this appeal, he replied that the President had seen fit to relieve him, and it would have so to be, unless the order was countermanded. Lieutenant Generals Hardee and Stewart then joined me in a telegram to the President, requesting that the order for his removal be postponed, at least till the fate of Atlanta was decided.

The following extract from a letter of Lieutenant General A. P. Stewart will show that I was desirous General Johnston should remain in command:

"ST. LOUIS, *August 7th, 1872.*

"GENERAL J. B. HOOD.

"MY DEAR GENERAL:—Your letter of the 25th ultimo was received some days since, and I avail myself of the first opportunity to answer it.

"You ask me to send you 'a statement setting forth the facts as you (I) understand them, of the circumstances attending the removal of General J. E. Johnston from the command of our Army in Georgia, in

1864, and my appointment to succeed him.' It gives me pleasure to comply with your request. * * * Monday morning, (July 18th,) you will remember we met about sunrise in the road near Johnston's headquarters; and I then informed you of the object of seeking an interview, and that was that we should all three unite in an effort to prevail on General Johnston to withhold the order, and retain command of the Army until the impending battle should have been fought. I can bear witness to the readiness with which you concurred. We went together to Johnston's quarters, and you and he had a long conversation with each other, which I did not hear. At the close of it, however, you and General Hardee and I went into the Adjutant General's office, and together prepared a telegram to the President, stating that, in our judgment, it was dangerous to change commanders at that juncture, and requesting him to recall the order removing Johnston, at least until the fate of Atlanta should be decided. That was the substance; I cannot remember the language. An answer was received that afternoon from the President, declining to comply with our request or suggestion, on the ground that the order having been issued, it would do more harm than good to recall or suspend it. * * *

"Very sincerely yours,

"ALEX. P. STEWART,

"Late Lieutenant General C. S. Army."

The President's answer to our telegram was as follows:

"RICHMOND, *July 18th, 1864.*

"TO GENERALS HOOD, HARDEE AND STEWART.

"Your telegram of this date received. A change of commanders, under existing circumstances, was regarded as so objectionable that I only accepted it as the alternative of continuing a policy which has proven disastrous. Reluctance to make the change induced me to send a telegram of inquiry to the Commanding General on the 16th inst. His reply but confirmed previous apprehensions. There can be but one question which you and I can entertain, that is, what will best promote the public good; and to each of you I confidently look for the sacrifice of every personal consideration in conflict with that object. The order has been executed, and I cannot suspend it without making the case worse than it was before the order was issued.

"JEFFERSON DAVIS."

After the receipt of the above telegram, I returned to General Johnston's room, alone, and urged him, for the good of the country, to pocket the correspondence, remain in command, and fight for Atlanta, as Sherman was at the very gates of the city. To this my second appeal he made about the same reply as in the first instance. I then referred to the great embarrassment of the position in which I had been placed; asserting, moreover, I did not even know the position of the two remaining corps of the Army. With all the earnestness of which man is capable, I besought him, if he would, under no circumstances retain command and fight the battle for Atlanta, to at least remain with me and give me the benefit of his counsel whilst I determined the issue. My earnest manner must have impressed him, since, with tears of emotion gathering in his eyes, he finally made me the promise that, after riding into Atlanta, he would return that same evening. Although our relations were, as they had been throughout the campaign, friendly and cordial, he not only failed to comply with his promise, but, without a word of explanation or apology, left that evening for Macon, Georgia.

CHAPTER VIII.

REPLY TO GENERAL JOHNSTON—HANDLING OF TROOPS—LEE AND JACKSON SCHOOL *versus* THE JOHNSTON SCHOOL—JOHNSTON'S PLAN TO HOLD ATLANTA "FOREVER."

GENERAL JOHNSTON makes the following arraignment:*

"General Hood asserts in his published report, that the Army had become demoralized when he was appointed to command it, and ascribes his invariable defeats partly to that cause. The allegation is disproved by the record of the admirable conduct of those troops on every occasion in which that General sent them to battle—and inevitable disaster. Their courage and discipline were unsubdued by the slaughter to which they were recklessly offered in the four attacks on the Federal Army near Atlanta, as they proved in the useless butchery at Franklin. He also states, †'It is a calumny to say that the Army of Tennessee was dispirited or broken down.' It had never before been in finer condition—the men in a high state of discipline and full of confidence from uniform success in their engagements with the enemy."

At the date of my transfer to the West, I, still under the influence of the teaching of Lee, Jackson, and Longstreet, could not but recognize a marked difference, after the crossing of the Chattahoochee river, between the troops of the Army of Tennessee and those of Virginia. My long experience and service with the latter, who formed, their limited numbers notwithstanding, one of the most powerful as well as renowned

* Johnston's Narrative, page 365.

† Johnston's Narrative, page 349.

Armies the world has produced, enabled me also to discover a marked difference in the spirit and *morale* of General Johnston's Army when south of the Chattahoochee, and when lying at Dalton, full of hope and anxious for battle. The cause of this difference is simple, and easily understood by those who have had a practical demonstration of the superiority of the Lee and Jackson manner of handling troops over the Joe Johnston mode of warfare. The one school elevates and inspirits, whilst the other depresses, paralyzes, and, in time, brings destruction. The effect of these respective schools is alike upon almost all men; otherwise some ground might exist for the assertion that the men of Lee's Army were of a superior class to those under Johnston. Not so, indeed. The *personnel* of the two Armies was originally of the same element, and there is no reason why our Army at Dalton, handled according to the Lee and Jackson school, should not have been made to equal its counterpart in Virginia. Although it may be argued that the Army of Tennessee had been dispirited, or demoralized, previous to its reorganization by General Johnston in the Spring of '64, it is nevertheless certain that, at the time of the first appearance of the Federals in its front at Dalton, it possessed the capability to be rendered the equal of the best troops in the Confederacy. In this assertion, I am confident I shall be upheld by the intelligent officers and men of that Army.

I regret to find it necessary to discuss this purely military question, since I have as warm personal friends in the Western as in the Virginia Army, and would be pained to know that aught from my pen had given umbrage to any Confederate, who performed his duty faithfully unto the end. I reiterate that the *personnel* of the two Armies was originally the same; that the troops at Dalton were capable of having been made the equal of those in Virginia. Therefore, I see not that two brothers—one having served in Lee's, and the other in Johnston's Army—have cause of jealousy, if one has accomplished somewhat more than the other; whereas had the two been under the same commander, they would have proved soldiers of equal merit. With these premises, I shall proceed to show in brief the cause of difference between these brother-soldiers of opposite schools.

General Lee never made use of entrenchments, except for the purpose of holding a part of his line with a small force, whilst he assailed the enemy with the main body of his Army—as, for instance, around Richmond at the time of the battle of Gaines's Mills—and save *en dernier resort*, as at Spottsylvania, to and around Petersburg, toward the close of the war. He well knew that the constant use of breastworks would teach his soldiers to look and depend upon such protection as an indispensable source of strength; would imperil that spirit of devil-me-care independence and self-reliance which was one of their secret sources of power, and would, finally, impair the *morale* of his Army. A soldier cannot fight for a period of one or two months constantly behind breastworks, with the training that he is equal to four or five of the enemy by reason of the security of his position, and then be expected to engage in pitched battle and prove as intrepid and impetuous as his brother who has been taught to rely solely upon his own valor. The latter, when ordered to charge and drive the enemy, will—or endeavor to—run over any obstacle he may encounter in his front; the former, on account of his undue appreciation of breastworks and distinct remembrance of the inculcations of his commanding officer, will be constantly on the look-out for such defences. His imagination will grow vivid under bullets and bombshells, and a brush-heap will so magnify itself in dimension as to induce him to believe that he is stopped by a wall ten feet high and a mile in length. The consequence of his troubled imagination is that, if too proud to run, he will lie down, incur almost equal disgrace, and prove himself nigh worthless in a pitched battle.

A somewhat similar result is to be observed in engagements, in the open field, with the red men of the forest. Those who are familiar with their mode of warfare well know that, whenever they are attacked away from such shelter as trees and boulders, they at once become confused, and scatter in all directions. I concede that five hundred, in the open field, would overpower one hundred men, howsoever well trained; but two hundred and fifty properly trained soldiers should always prove the equal of five hundred Indians, mainly because of the difference in the manner of handling forces, practiced by the respective combatants. On the one hand, shelter is invariably sought in time of battle; on the other, reliance is placed upon boldness and valor.

In accordance with the same principle, a cavalryman *proper* cannot be trained to fight, one day, mounted, the next, dismounted, and then be expected to charge with the impetuosity of one who has been educated in the belief that it is an easy matter to ride over infantry and artillery, and drive them from the field. He who fights alternately mounted and dismounted, can never become an excellent soldier of either infantry or cavalry proper. Moreover, the highest perfection in the education of troops, well drilled and disciplined, can only be attained through continued appeals to their pride, and through incitement to make known their prowess by the substantial test of guns and colors, captured upon the field of battle. Soldiers thus educated will ever prove a terror to the foe. The continued use of breastworks during a campaign, renders troops timid in pitched battle; and the employment of such defences is judicious and profitable alone when resorted to at the proper time. They should be used not unto excess, and only in such instances as I have already mentioned, and in such as I shall hereafter specify. The result of training soldiers to rely upon their own courage, we behold in the achievements of Lee's troops. Long will live the memory of their heroic attempt to scale the rugged heights of Gettysburg; of their gallant charge over the breastworks at Gaines's Mills, and again over the abatis and strong entrenchments at Chancellorsville; of the many deeds of equal daring, which history will immortalize.

I shall consider, for a moment, the manner in which General Lee handled his troops. After the battle of Sharpsburg, or Antietam, McClellan followed him south of the Potomac; instead of forming line of battle, and throwing up entrenchments upon every suitable hill he could find, from Maryland to the Rapidan, for the purpose of skirmishing, and delaying the enemy—which work he properly left to the cavalry—he threw his colors to the breeze, and, with martial music, marched to the line of Gordonsville and Fredericksburg. A few months later, when the Federals appeared in his front, he marshaled his forces, which, refreshed by their long rest, were anxious for battle; he at once attacked, defeated the enemy, and pursued him to the Potomac. He thus drove back, successively, Pope, Burnside, and Hooker.

After the battle of Gettysburg, Meade likewise followed Lee south of the Potomac. Again, he marched to the line of the Rapidan, as in the first instance, leaving his cavalry to observe and check the advance of the enemy. General Grant subsequently appeared in his front, with a large and well-equipped Army. Although our great chieftain had only about forty-five thousand (45,000) effective men wherewith to oppose him, he, true to his past history, attacked instantly—having cut roads through the Wilderness, in order to get at the enemy—and so fierce was his assault that it almost made the very stones of the earth cry out. History will relate how nigh he was, in this instance, unto the achievement of victory; so nigh, indeed, that Mr. Lincoln, if I remember correctly, remarked in a speech in the course of which he referred to this desperate onslaught, that Grant had been jostled, not driven back; and that any one of the men he had sent previously to the command of the Army of the Potomac, would have been back on the north side of the Rappahannock.

Thus it will be seen that General Lee made use of entrenchments only *en dernier resort*, as around Petersburg, or in order to hold one portion of his line with a small force whilst he attacked with the main body; also that when he found it necessary to retreat, or fall back from an advanced position, he marched his Army to the line he intended to defend, instead of constantly fighting, skirmishing, avoiding a general engagement, and taking up position, day after day, to be abandoned under cover of darkness. General Johnston not only made uniform use of entrenchments, but retreated and fought at the same time—an error which Lee carefully eschewed, and one which should always be avoided, since the long continuance of such policy will prove the inevitable ruin of any army. Napier, one of the highest authorities on war, says: "It is *unquestionable* that a *retreating army* should fight *as little as possible*." Such was, however, the mistake committed by General Johnston. If he did not intend to risk a battle in the mountain fastnesses between Dalton and the Chattahoochee, but preferred to decide the fate of Georgia, the centre of the Confederacy, upon the flat plains around Atlanta, he should have left the cavalry in his rear to check the advance of the enemy; have marched his Army direct to the latter point, without firing a musket; and there have awaited Sherman's advance, when he should have made his attack. By the pursuance of this policy, he would have been able to engage Sherman with over seventy thousand (70,000) effective men, instead of fifty thousand (50,000) he claims to have had after crossing the Chattahoochee river. In lieu thereof, a course was pursued which entailed a loss of twenty-five thousand (25,000) men, without a single general battle having been fought, and which seriously demoralized the next to the largest and proudest Army assembled in the South.

When I state the Army was demoralized, I desire, at the same time, to except not only men who performed individual acts of remarkable devotion and courage, but also brigades and divisions, which, in prowess and discipline, would compare with the best troops in any army; unfortunately, however, the efforts of one such brigade or division were paralyzed by others so thoroughly effected by their training in the

Johnston school as to render them of but little service in a pitched battle.

A policy similar to that of my predecessor can be persisted in till desertions will take place by the thousands. The longer an army retreats, entrenches, and fights at the same time, the more numerous the desertions, and the more thorough the demoralization. As I have already mentioned, Lee handled his troops upon a directly opposite basis. They were always taught to work out the best means to get at the enemy, in order to cripple or destroy him, in lieu of ever seeking the best means to get away from him. Therefore the Lee and Jackson school is the opposite of the Joe Johnston school, and one will always elevate and inspirit, whilst the other will depress and paralyze.

The statement of Lieutenant Generals Hardee and Stewart, to the effect that the Confederate Army, after crossing the Chattahoochee, had as much spirit and confidence as it possessed at Dalton, is erroneous. Whilst I have a proper regard for the opinions of these officers who spoke, I believe, in all sincerity, I cannot but consider that their impressions were formed from their own standpoint, without having actual knowledge of the high state of perfection obtained by the troops in the Virginia Army, under the training and mode of handling of General Lee. In the course of daily life our thoughts and convictions generally receive their impress from our surroundings; and, if we confine our experiences to any one sphere of life, without contact with the various spheres around us, we lose that power of comparison by which we are enabled to form correct judgments of things and men. These officers formed their decision from but one standpoint, which was the Army of Tennessee, and they comprehended not fully the spirit of heroism which pervaded the Army led by our great chieftain to victory after victory. Therefore they were partial judges when came into question the comparative spiritlessness of the Western Army, as it slowly retreated a distance of one hundred miles, without a single glorious victory to inscribe upon its banners.

If requisite, I could bring forth abundant evidence from officers of that Army that the continuous retreat from Dalton to the plains of Georgia, produced a demoralizing effect. General Frank Blair, whose corps was engaged in the battle around Atlanta on the 22d of July, 1864, when my friend and classmate, General McPherson, was killed, states in a letter to a prominent officer of the Army of Tennessee, that the Confederate troops, on that day, did not fight with the spirit they should have displayed. It was, nevertheless, reported to me, at the time of this engagement, that they had fought with gallantry, and I so telegraphed to the authorities at Richmond. The truth is, no troops handled as these had been from Dalton to Atlanta could have attacked with extraordinary vigor, and I do assert that fifty thousand men of the Lee and Jackson school will always prove equal to eighty thousand (80,000) of the Johnston school; moreover, that the small Army I commanded at Franklin was equal to that which was turned over to me at Atlanta, although it numbered only about one-half in effective strength, for the simple reason that a forward march of about one hundred and eighty miles, together with a different mode of handling it, had contributed to the improvement of its *morale* and the restoration of its pristine spirit. These conclusions I have reached after a long and careful consideration of the subject. It has been my fortune to serve, during the war, in every grade from that of First Lieutenant to that of Commander-in-Chief. Having, therefore, been under fire with both small and large bodies of men, and having carefully observed the effect of such fire upon troops with and without breastworks, the principles which I have endeavored to elucidate will, in my opinion, stand the test of time.

In January, 1874, I addressed the following communication to Lieutenant General Stephen D. Lee, who served a long period in Virginia, and subsequently in the Army of Tennessee:

"NEW ORLEANS, *January 17th, 1874.*

"GENERAL:—Your position during the late war, and experience throughout different campaigns in Virginia, doubtless enabled you to observe, and form an opinion of the general effect of entrenchments upon an army. Since the close of the revolution I have conversed with many officers of the Army of Northern Virginia upon this subject, and have been informed that when General Lee was forced, as a *dernier* resort, to use breastworks around Petersburg, it had a depressing effect even upon the stern veterans who made up that grand old Army; that it could easily be discerned when the troops were called upon to leave the trenches, and again give battle in the open field. During three years' service in the Virginia Army, as regimental, brigade, and division commander, under the orders of Generals Lee, Jackson, and Longstreet, I was never required to throw up even temporary breastworks for the protection of my troops. The battles of Gaines's Mills, Second Manassas, Fredericksburg, Sharpsburg, Chancellorsville, and Gettysburg, were all fought by the Confederates without the aid of such defences. The officers and soldiers, who served in the Virginia Army, know of the great self-reliance and spirit of invincibility which pervaded its ranks, and how correct the appreciation of General Lee, when he said, 'There were never such men in an army before; they will go anywhere, and do anything if properly led.' Those who come after us will seek the cause of the extraordinary results accomplished by the Army of Northern Virginia in comparison with other Armies of the South. The *personnel* of all the Confederate Armies being about the same, the question must arise, and will be discussed, as to whether there was not something in the handling of the troops or in the strategy and tactics, adopted and carried out by the distinguished leaders of Virginia, which produced soldiers equal to twice the number of the enemy.

"I of course admit the necessity of fortifications for the protection of certain harbors, depots, and important centres, which should, however, never be allowed to become pitfalls for large bodies, but be well provisioned, and garrisoned only by a sufficient number of men to stand a siege, if requisite; and whilst I also admit the necessity of entrenchments under such circumstances of constraint as those which induced General Lee finally to resort to them, my own experience has taught me that the continued use thereof, by an army has a demoralizing influence.

"An army cannot at one time fight behind breastworks, with the practical demonstration that its position renders it equal to three times the strength of the foe, and, at another time, when occurs a favorable opportunity of attack, forget its own experience in the use of entrenchments, and charge the enemy's works, and force him to fly from the field, as was so often the case in Virginia.

"Hoping soon to have your valued opinion upon this subject,

"I am truly yours, J. B. HOOD."

I received the subjoined in reply:

"BROOKEVILLE, MISSISSIPPI, *January 26th, 1874.*

"TO GENERAL J. B. HOOD.

"GENERAL:—In your favor of the 17th inst., you ask my opinion of the 'general effect of entrenchments upon an Army.' My experience during the recent war was nearly equally divided in serving with and without entrenchments.

"My service with the Army of North Virginia ended after the battle of Sharpsburg—then in the campaigns in Mississippi, involving the fall of Vicksburg—again in the campaign in Georgia, involving the fall of Atlanta, and also the last campaign into Tennessee. Entrenchments were generally used in my service in the West. They were not used in Virginia up to the time I was transferred West. I am free to say that I consider it a great misfortune to any army to have to resort to entrenchments; its *morale* is necessarily impaired from their constant use.

"Troops once sheltered from fire behind works, never feel comfortable unless in them. The security of entrenchments is a constant subject of discussion by troops who use them. It is a matter of education. They are taught that one man equals five or six of the enemy. This they remember when called upon to attack entrenchments of the enemy, about which they are necessarily timid. Troops in works, engaged the first time, are always bolder than afterwards—stand erect and deliver their fire with precision as they were used to in the open field; after a few engagements, the thought of constant security is always with them, and their object is to be always covered by the works, while under fire.

"An army, accustomed to entrenchments, has its efficiency impaired as a whole, from the fact that in nearly every division one or two brigades consider it hazardous in the extreme to attack entrenchments; hence, in the attack hesitate, and hesitancy in attacking works is certain defeat. A bold and defiant attack on works, though attended with great loss when successful, generally drives troops from the works before reaching them—which shows that boldness in attack and nearness make a sudden change in the ideas of the troops behind the works, they being discouraged, and disappointed in not seeing the enemy easily and certainly driven back.

"A general who resorts to entrenchments, when there is any chance of success in engaging in the open field, commits a great error. Entrenchments are sometimes necessary for the safety of an army, encountering greatly superior numbers to gain time, or to save an army defeated by superior numbers. When these occasions are plainly visible to the army, I do not see that its efficiency is necessarily impaired, when encountering the enemy again in the open field; but the habitual use of entrenchment certainly impairs the boldness of attack in any army.

"To attack entrenchments, give me troops who have never served behind them. Good troops, in line of battle, before using entrenchments, feel as secure without works as with them. As an instance, recall your division, at Sharpsburg, when attacked by more than five times its number in an open field, or again your brigade at Gaines's Mills, when it carried the works of the enemy.

"I would not be understood as arguing against the use of entrenchments when the occasion is plain for their use; but certainly against an army habitually using them; for the latter use of them destroys a plan of campaign, and there is no campaign; and the generals are besieging, or resisting a siege in fact, and with casualties from day to day soon equivalent to the loss in a general engagement.

"I am yours, truly,

"STEPHEN D. LEE."

This officer enjoyed a fair opportunity to note the effect of long continued use of entrenchments upon an army. He served with distinction until the close of the war, and displayed superior ability as a corps commander whilst in the Army of Tennessee.

A forcible example of the difference in the mode of handling troops—as illustrated by General Lee, and those generals who constantly resort to the use of fortifications—is afforded in the recent war between France and Prussia, and also in that between Turkey and Russia. One hundred and fifty thousand of Louis Napoleon's Army, under Bazaine, shut themselves up in the stronghold of Metz; allowed a much larger number of Germans to surround them; to construct works almost as formidable as their own, and quietly await their surrender for want of provisions.

The Turks committed a similar blunder at Plevna, in allowing a fine army to be entrapped and the remainder of their forces to become demoralized by the natural effect of this gross error, in lieu of holding their troops well in hand, taking some general line upon which to retard and cripple the enemy as much as possible, and, finally, beat him, if not in a general pitched battle, in detail; the only chance of success for the weaker power grappling with its strong enemy. Had

General Lee pursued the plan of the French or of the Turks, he would have entrenched himself at an early day of the war —say at Winchester; have allowed the Federals to surround him with five times his numbers; to construct breastworks, finally, to compel him to surrender, and thus bring demoralization to his countrymen, in addition to the loss of their cause.

The "reckless" attacks around Atlanta—so designated by General Johnston—enabled us to hold that city forty-six days, whereas, he abandoned in sixty-six days one hundred miles of territory, and demoralized the Army.

It is a significant fact that General Sherman dedicates only thirty-eight pages to an account of the rapidity with which he dislodged Johnston from one position after another in the mountains from Dalton to Atlanta, and devotes that number to an explanation of the necessary operations of his Army, in order to force me to abandon the *one* untenable position of Atlanta.

General Johnston says: *

> "General Hartsuff, General Schofield's Inspector General, told me, in the succeeding Spring, that the valor and discipline of our troops at Franklin, won the highest admiration in the Federal Army."

The valor displayed at Franklin, and which deservedly won the admiration of the Federals, was caused by the handling of the troops in a directly opposite manner to that of General Johnston, together with the advance movement previously inaugurated, and the mortification experienced after the unfortunate failure the day before at Spring Hill. Inasmuch as General Johnston never inaugurated a forward movement, nor sought out the enemy, but invariably retreated in their front, he is not able to comprehend the origin of the gallantry so conspicuous on that field. He, therefore, errs as egregiously in the supposition that his continued retreat from Dalton to Atlanta and incessant entrenching gave rise to the courage displayed, at Franklin, as in his endeavor to find a parallel to

* Johnston's Narrative, page 365.

his campaign in that of Lee against Grant, from the Rappahannock to Petersburg: they in truth are the opposite of one another.

General Johnston states,* "In transferring the command to General Hood I explained my plans to him." He may have said somewhat to me in regard to his plans—if, indeed, he had at any time resolved upon the defence of Atlanta—but I have no recollection thereof; possibly, from the fact that I was thoroughly engrossed by the grave responsibilities unexpectedly thrust upon me at that critical moment. At all events, we are now informed, through his Narrative, that he had two plans, and that, if the first had failed, the second would, at least, have secured to the Confederacy Atlanta "forever."

Thus would have been wrought our independence, and the Southern people have been spared the sorrow and degradation to which they were so long subjected. If General Johnston be correct in his assertion that no reason exists why Atlanta should not have been held "forever," a heavy responsibility rests upon the Confederate authorities who relieved him of the command of the Army of Tennessee.

Heavier still is the responsibility assumed by them, when they refused to dismiss General Lee from the command of the Army of Northern Virginia, and to re-assign General Johnston to that position, after his recovery from a wound received at the battle of Seven Pines.

He states, in addition,† that his "Army had a place of refuge in Atlanta, too strong to be taken by assault, and too extensive to be invested." According to his theory, Richmond, which was larger than Atlanta, should also have been too extensive to be invested; and its defences, which I am certain any council of competent officers would pronounce more tenable than those of its sister city, should also have been too strong to be carried by assault. It follows, therefore, that if General Johnston could have held Atlanta "for-

* Johnston's Narrative, page 350.

† Johnston's Narrative, page 358.

ever," most assuredly would he have held Richmond "forever," and have given us that freedom for which the great Lee struggled so gloriously, but in vain. Again, if this General felt it within his power to hold Atlanta "forever," unpardonable is the offence he committed, in refusing to answer definitely, when interrogated by the President who was anxious to ascertain whether or not he intended to defend Atlanta. In view of the abandonment of one hundred miles of territory into the very heart of the country, it was but natural the Government should have made such inquiry; and who, with any degree of justice, can question the right of the authorities, at Richmond, to have sought, nay demanded, a positive answer from one of their subordinate officers? Had General Lee been placed in the same position, how long would he have hesitated to answer most fully and satisfactorily the President's inquiry on the 16th of July?

If General Johnston had, at that time, informed President Davis that he could see no reason why Atlanta should not be held "forever," he would have been retained in command. I know this to be true; moreover, the correspondence I have already published, clearly indicates this fact. Lastly, if his declaration in regard to the tenability of Atlanta be grounded upon sound principles, who of my countrymen will forgive him for having deserted me under the peculiar trials of the hour, instead of aiding me by his counsels to accomplish the great end, at the sacrifice of every personal consideration, and in the spirit of a true patriot.

These are, indeed, grave questions, and afford matter for serious reflection to every Southerner, especially since General Johnston claims, by asserting his ability to have held Atlanta "forever," the power to have saved the Confederacy from the disaster and ruin which followed. As already stated, the order relieving him from the command of the Army was received upon the 17th of July, at 11 p. m., he, unwilling to await the dawn of day, promulgated the order that night to the troops, and by dark, the next evening, he was journeying towards Macon with all speed possible. Had he remained with the Army, at my urgent solicitation, he would undoubtedly, have gained the credit of saving Atlanta, in the event of success; in case of failure, his friends could, as they have already done, have taken measures to protect his reputation by asserting that I had not altogether followed his counsels. The responsibility of non-success would have rested upon me, whilst he had nothing to lose, and all to gain. He was, however, in so great haste to leave the scene of action that I have almost been inclined to think he was rejoiced at having been relieved from the duty of holding Atlanta "forever," and thus insuring the independence of his people.

It now devolves upon me, in order to vindicate myself, as well as the Confederate Government, and to exonerate my predecessor from the charge of apparent insensibility to the fate of Atlanta and his country, to show that some doubt actually exists of his ability to have held Atlanta "forever."

First, I will consider the evidence to be presented against his intention, at any time, to fight for this city, and then demonstrate the insufficiency of his power to make good an assertion which, after an interim of nigh ten years, is, for the first time, published to the world.

It was generally believed, before the Army abandoned Dalton, that General Johnston would make a stand at that point; throughout his correspondence with the Government, during the Winter and Spring of 1864, and in which he urges all available troops to be sent immediately to his command, one is led to suppose that he actually intended to fight at that stronghold. In his letter to President Davis, dated January 2d, 1864, he speaks thus:* "I can see no other mode of taking the offensive here than to beat the enemy when he advances, and then move forward." In response to General Bragg's letter of March 12th, proffering fully eighty thousand (80,000) men, as an inducement to assume the offensive, and to which

* Johnston's Narrative, page 275.

letter I have already referred, General Johnston dispatched the following telegram: *

"Your letter by Colonel Sale received. Grant is at Nashville. Where Grant is we must expect the great Federal effort. We ought, therefore, to be prepared to beat him here"—at Dalton. In his written reply to the same, he says: † "We cannot estimate the time he (the enemy) will require for preparation, and should, consequently, put ourselves in condition for successful resistance as soon as possible by assembling here the troops you enumerate." Again, ‡ "I would have the troops assemble here without delay, to repel Grant's attack and then make our own."

It is hereby evident that as long as General Johnston endeavored to obtain the transfer, to his own command, of Longstreet's Corps in Virginia, and of Polk's Army in Mississippi, he spoke continually of fighting at Dalton; when, however, Sherman appeared at Tunnel Hill, in front of Rocky-faced Ridge, and he was given an Army of over seventy thousand (70,000) available troops—as I have demonstrated—he decided to retreat. What followed at Resaca? Retreat. New Hope Church? Retreat. Cassville? Retreat. Kennesaw Mountain? Retreat. Would we have fought at Atlanta after our inglorious campaign, the abandonment of the mountain fastnesses, and the foreshadowed intention of our commander to fall back to Macon?

I shall now glance at his two plans for the defence of Atlanta, one of which was to insure the security of that city "forever." By his first plan, he hoped to attack the enemy as they crossed Peach Tree creek. Within thirty-six hours, almost before he had time to select quarters in Macon after his departure on the evening of the 18th of July, General Thomas was crossing Peach Tree creek, whilst McPherson and Schofield were moving to destroy the railroad to Augusta. General Johnston evidently had little faith in this plan, since he was unwilling to await thirty-six hours to test its feasibility.

---

* Johnston's Narrative, page 294. † Johnston's Narrative, page 295.
‡ Johnston's Narrative, page 296.

By his second, and, "far more promising plan," as he designates it, he intended to man the works of Atlanta, on the side towards Peach Tree creek, with the Georgia State troops; and, upon the approach of the enemy, to attack with the three corps of the Army in conjunction with the cavalry.

When the advance sheets of Johnston's Narrative appeared before the public, I read with amazement the account of this extraordinary project, and, forthwith, addressed the following letter to Major General Gustavus W. Smith, who commanded the Georgia State troops previous to General Johnston's removal, and during the siege of Atlanta:

"NEW ORLEANS, *January 17th, 1874.*

"GENERAL G. W. SMITH, *Frankfort, Ky.*

"GENERAL:—Having occasion to refer to your official report of the operations of the Georgia Militia around Atlanta, I find you were assigned the command of these troops the 1st of June, 1864; that you relieved General Wayne, who had been placed under, and subject to, the orders of General Johnston; that you had, at the time you crossed the Chattahoochee and marched to the support of General Johnston, about three thousand (3000) effective men; that when you recrossed the river and reported to me for duty in the trenches around Atlanta, you had about two thousand (2000) effective muskets. You also state that Governor Brown called out the reserves; that the largest number of effective men in your command, at any one time during the siege, did not exceed five thousand (5000); that they were poorly equipped, more than two-thirds of them having no cartridge boxes; that most of the reserves had never been drilled and the others but a few days, etc., etc.

"Your relations with the Governor of Georgia, at this time, were such as to enable you to form a correct opinion as to whether or not Governor Brown furnished me, for the defence of Atlanta, as many State troops as he could or would have furnished General Johnston, had the latter remained in command. The Governor gave me to understand that he had sent forward all the militia he could enroll, and I have no reason to doubt his assertion.

"Your large experience as a soldier, moreover, enables you to form an estimate as to the ability of five thousand (5000) militia to have occupied the trenches in front of the enemy, and have held Atlanta against General Sherman's Army of over one hundred thousand (100,000) effective men, and thereby to have rendered free the three corps, which constituted the whole Army of Tennessee, and have allowed them to operate on the outside against either flank of the enemy. My reason for requesting your military opinion upon this subject is that in General Jos. E. Johnston's forthcoming book appears the following statement:*

"'In transferring the command to General Hood I explained my plans to him. First, I expected an opportunity to engage the enemy on terms of advantage while they were divided in crossing Peach Tree creek, trusting to General Wheeler's vigilance for the necessary information. If successful, the great divergence of the Federal line of retreat from the direct route available to us would enable us to secure decisive results; if unsuccessful, we had a safe place of refuge in our entrenched lines close at hand. Holding it, we could certainly keep back the enemy, as at New Hope Church and in front of Marietta, until the State troops promised by Governor Brown were assembled. Then, I intended to man the works of Atlanta on the side toward Peach Tree creek with those troops, and leisurely fall back with the Confederate troops into the town, and, when the Federal Army approached, march out with the three corps against one of its flanks. If we were successful, the enemy would be driven against the Chattahoochee where there are no fords, or to the east, away from their communications, as the attack might fall on their right or left. If unsuccessful, the Confederate Army had a near and secure place of refuge in Atlanta, which it could hold forever, and so win the campaign, of which that place was the object. The passage of Peach Tree creek may not have given an opportunity to attack; but there is no reason to think that the second and far most promising plan might not have been executed.'

"Whilst I acknowledge with pleasure the gallant conduct and efficient service of the Georgia State troops in the defence of Atlanta, I cannot conceive how they could have been expected to accomplish all that General Johnston seems to have anticipated, *i. e.*, man so long a line of breastworks as that on the side of Peach Tree creek, which embraced the front of General Sherman's entire Army; and when, as you will remember, within three days after General Johnston relinquished the command, the enemy's left was across the Augusta Railroad, southeast of the town, and moving rapidly southwest to destroy the railroad to Macon.

"Your views upon this important subject, I should be pleased to have at your earliest convenience.

"Yours truly,
"J. B. HOOD."

---

* Johnston's Narrative, pages 350, 351.

"FRANKFORT, KENTUCKY,
" *January 23d, 1874.*

"GENERAL JOHN B. HOOD, *New Orleans, La.*

"GENERAL:—Your letter of the 17th inst. is received. In answer to your first inquiry I have to say that, in my opinion, you were furnished with all the State forces that the Governor of Georgia, could by the use of extraordinary powers bring to assist in the defence of Atlanta.

"Your second question calls for my opinion 'as to the ability of five thousand (5000) militia to have occupied the trenches in front of the enemy and have held Atlanta against General Sherman's Army of over one hundred thousand (100,000) effective men, and thereby to have rendered free the three corps, which constituted the whole Army of Tennessee, and have allowed them to operate on the outside against either flank of the enemy.'

"Atlanta would, in all probability, have been taken by the enemy within twenty-four hours after its defence was entrusted to the Georgia militia, because in number this force was entirely inadequate, under the circumstances.

"Very truly yours,

"GUSTAVUS W. SMITH."

I was unwilling to harbor a suspicion that Governor Brown would have furnished for the defence of the State, and of our common cause, a larger number of troops to General Johnston than to myself; neither could I perceive in what manner the impossibilities, suggested by this General, were to be accomplished by the Georgia militia. General Gustavus W. Smith is a soldier, as well as an engineer, of eminent ability, and his opinion is entitled to much weight in a discussion of this character.

The side towards Peach Tree creek embraced about the entire front of General Sherman's Army. I, therefore, found it necessary to place not only the Georgia State troops on that side, but also two corps of the Army, whilst I made the attack of the 22d of July with a single corps and the cavalry. I very much regret General Johnston's inability to have remained, and enlightened me in regard to the means to hold Sherman's one hundred and six thousand (106,000) at bay with five thousand (5000) militia, whilst I attacked one of the enemy's flanks with the entire Army of Tennessee. If this feat could have been achieved, great results might have ensued. In view of General Johnston's now avowed intention to have made a stand at Atlanta, it would certainly have been more judicious to have marched direct to the line he had resolved to defend—as General Lee marched out of Maryland and Pennsylvania to the Rapidan; to have thus reserved the twenty-five thousand (25,000) effective men and nineteen thousand (19,000) muskets, lost on our retreat through Georgia, and have used these trained soldiers and good muskets on the side towards Peach Tree creek, instead of General Gustavus W. Smith's five thousand militia, many of whom were armed with flint lock muskets, and were devoid even of cartridge boxes. Surely this plan would seem to have been more feasible, and certainly more promising.

If Sherman had not a sufficient force to form a cordon of troops round the city, he was able to accomplish his object by equally effective means. The size of Atlanta in no manner hindered the destruction of our railway line of communications which, in the exhausted condition of our resources the last year of the war, we were no wise competent to re-establish when great damage had been committed. We had neither the material nor the force to repair them.

If General Johnston considered Atlanta so especially adapted to his purposes, inasmuch as it was too extensive to be invested and too strong to be carried by assault, I am at a loss to divine the reason why he did not take a radius equal to that of Atlanta, and describe a circle from a given centre, within the mountains of Georgia; throw up entrenchments, and declare to the world the impregnability of his position and his intention to hold it "forever." Trees of the forest would certainly have been of as much or more service to him than buildings proved to me in Atlanta, or to General Lee in Petersburg, and Richmond.

No more decided advantage was to be derived from the junction of railways, at Atlanta, than was afforded at Kingston, or any other point on the railway line below Dalton, because of our poverty in resources towards the close of hostilities and consequent inability to reconstruct at the same time two or three roads when seriously damaged.

CHAPTER IX.

REPLY TO GENERAL JOHNSTON—HIS INTENTION TO ABANDON ATLANTA—EVACUATION OF RICHMOND CONTEMPLATED IN 1862—ATTEMPT TO COURT MARTIAL.

After General Johnston's abandonment of the mountains of Georgia, his inconsistency in maintaining that Atlanta was a position "too strong to be carried by assault," must indeed strike with surprise not only military men, but civilians of intelligence. From the earliest periods of history mountain fastnesses have proved, by actual test, the most secure strongholds to every people determined upon obstinate resistance. General Lee asserted shortly before the close of the war that he could continue the struggle twenty years longer, if hostilities were transferred to the mountain regions of the Confederacy. Numerous instances are adducible to attest the fact that prolongation of war is ofttimes to be attributed to such protection from Nature. To cite an example from a neighboring island, how long, I venture to inquire, would the Cubans have held out against the Spanish Government, if their territory had been devoid of mountains, and had stretched forth to the sea in one vast plain like the country from Atlanta?

But when the Confederate commander, with seventy thousand available men, surrendered the Thermopylæ of the South without risking a general battle, it is hardly reasonable to suppose that he would have made a final stand upon the plains of Georgia.

According to the following extract from an official telegram, even General Sherman was in doubt as to whether or not Johnston would fight for Atlanta: *

"HEADQUARTERS MILITARY DIVISION OF THE MISSISSIPPI,
"IN THE FIELD, AT SAN HOUSE, PEACH TREE ROAD,
"FIVE MILES N. E. OF BUCKHEAD, GA., *July 18, 1864.*

* * * "It is hard to realize that Johnston will give up Atlanta without a fight, but it may be so. Let us develop the truth.

"W. T. SHERMAN,
"Major General Commanding."

My predecessor had evidently another scheme in reserve. General Forrest was required, with five thousand (5000) cavalry in Tennessee, to destroy Sherman's communications with Nashville,—at least, in so far as to hinder Sherman from receiving sufficient supplies for the maintenance of his Army. General Wheeler's cavalry force numbered over ten thousand (10,000), and was composed of as brave men as those under the command of Forrest. If this force, with the exception of a small detachment to protect the flanks of the Army, was unable to break the Federal line of communications, I cannot conceive in what manner General Forrest was expected to accomplish this object with only five thousand (5000) men—especially, when Sherman had a large force of cavalry attached to his own Army, as well as another large body of this arm in Tennessee; had erected block houses at every important bridge and culvert, and had stationed infantry at fixed points along the entire line between Nashville and Atlanta, forming, it might be said, a chain of sentinels. The Federals had at their disposal locomotives of great power, and a sufficient number of cars to move, within a few hours, a corps of infantry to any one threatened point. Their vast resources enabled

* Van Horne, 11 Army C., vol. II, page 121.

them also to rebuild the railroad almost as fast as Forrest could have destroyed it. General Johnston, therefore, errs in the supposition that five thousand cavalry, under these circumstances, could have so effectually destroyed Sherman's communications as to compel him to retreat. The impossibility of the success of this plan, however, will be clearly established when I give an account of the inability, during the siege of Atlanta, of Forrest's cavalry together with about five thousand under Wheeler to accomplish this important object.

I am, therefore, reluctant to believe that General Johnston possessed any more definite idea of defending Atlanta than he had of defending Dalton, or any other position from that point to Atlanta. He brings forward the presence of his family in this city, as evidence of his intention to make a stand; and affirms that the entrenchments thrown up, together with the moving forward of heavy artillery, support his testimony. Unfortunately, in view of his history in the past, the evidence is not conclusive. He threw up various lines of works during his campaign, and, successively, abandoned them; moreover, whatever heavy artillery had been ordered to the front could, if the necessity had arisen, have been placed upon cars, and been removed to the rear. In regard to the first plea, I am unable to discover why his family could not retreat as well as the Army.

A General who, at New Hope Church, informed his corps commanders that he considered Macon, one hundred miles beyond Atlanta, the point to fall back upon, would hardly have resisted the temptation to carry out his suggestion, when to retreat was, with him, if not a fixed principle, certainly an inveterate habit.

Aside from any other evidence, the following extract from a letter received from General M. C. Butler, * now United States Senator, is sufficient to prove that General Johnston had no hope or idea of holding Atlanta.

---

* 1879.

"EDGEFIELD, S. C., *July 18th, 1874.*

* * * * * * "I was with General Johnston when he arranged the terms of surrender with Sherman. Generals Hampton and Wheeler being away at the time, I commanded the cavalry of Johnston's Army and accompanied him with an escort to the last interview with Sherman, and on our return to camp he told me that he had had no confidence in the success of our cause for two years." * * * *

As stated, if this General could have held Atlanta "forever," he likewise would have held Richmond "forever."

In this connection I will,—in defence of General Lee, make known an historical fact of singular interest, and of which I have but recently been apprised; it is true the matter was hinted about at the date of the occurrence, but I now, for the first time, receive the information from the highest authority.

About the 26th of April, 1874, I met, in Mobile, the Honorable C. M. Conrad, of Louisiana. We were each en route to New Orleans, and in the freedom of friendly conversation, we discussed without restraint the subject of the late war. General Johnston's book was referred to, when Mr. Conrad remarked that Mr. McFarland, of Richmond, Virginia, a volunteer aid on the staff of General Johnston at the time of his retreat from Yorktown—had informed him, during the war, that General Johnston said to him (Mr. McFarland), on the retreat from Yorktown, that he (Johnston) expected or intended to give up Richmond. Mr. McFarland expostulated and protested; finally expressed to the Commanding General the hope that he would change his mind. I at once observed to Mr. Conrad that this fact was truly an important link in the history of that period, and, if no objection existed upon his part, I might on some occasion refer to the incident. He replied, "Well, it is a matter of history," or words to that effect.

The above is almost verbatim the statement of Mr. Conrad to me, in Mobile. When we remember the high character of the late Mr. McFarland, a banker of Richmond, a citizen who was not only beloved and respected in Virginia, but well known to all the prominent men of the South as a gentleman of honor and unimpeachable integrity, and when we consider the name, the position, and the career of the Honorable C. M. Conrad, the testimony becomes irrefutable.

Richmond would have been abandoned by General Johnston at the outset of the struggle, had he been afforded the opportunity; in other words, had he not, in consequence of his disability, been replaced by General Lee, who retained, to the end, command of the Army of Northern Virginia.

Shortly after my return to New Orleans I resolved to obtain from Mr. Conrad a written statement of this important fact. He had, however, left for Washington. Thereupon I addressed him the following letter:

"NEW ORLEANS, *May 19th, 1874.*

"HONORABLE C. M. CONRAD, *Washington, D. C.*

"MY DEAR SIR:—I called at your office this morning to ask that you give me a memorandum of the statement of Mr. McFarland to you, in regard to General Johnston's giving up the city of Richmond at the time of his retreat from Yorktown. Since Mr. McFarland was, at this time, a volunteer aid of General Johnston, and was so well and so favorably known throughout Virginia, and by our prominent men of the South, any statement of his to one of your prominence in the public affairs of this country, makes a very important link in history—in fact becomes of great historical value. If no objection on your part, I would be much pleased to have you give me, in brief, what you stated to me on this subject, in Mobile, about the 26th ultimo.

"Respectfully and truly yours,

"J. B. HOOD."

I received in answer the following:

"WASHINGTON, D. C., *June 13th, 1874.*

"DEAR GENERAL:—On my return a few days ago from a visit to West Virginia, I found your letter and telegram. Upon reflection, I have determined that I cannot with propriety comply with your request.

"In the first place, although the conversation between Mr. McFarland and myself, of which you ask me to give you a statement, was not professedly confidential, it might, from its nature, be considered impliedly so, as it related to a communication which was confidentially made by General Johnston to him, and which he would, probably, not have divulged to any one but an intimate friend. It was perhaps indiscreet in me to have repeated the remark at all; but to give publicity to it, and that for a purpose unfriendly to General Johnston, would, in my view, be unjust both to General Johnston and to Mr. McFarland, as the latter is no longer alive to explain it, if necessary to maintain his statement.

"I will add that I have long known and esteemed General Johnston and his family. In his quarrel with Mr. Davis (which you are aware commenced long before the events which gave rise to the controversy between him and yourself), he had my sympathy and support. Under these circumstances for me to volunteer a statement not for the purpose of vindicating your military reputation, but for the purpose of assailing him in a matter with which you were in no manner concerned, would, I think, give just cause of complaint to him and his friends. I am sure that on considering the matter you will come to the same conclusion.

"Very truly yours,

"C. M. CONRAD."

I subjoin my reply:

"NIAGARA, ONTARIO, *July 1st, 1874.*

"HONORABLE C. M. CONRAD, *New Orleans, La.*

"DEAR SIR:—I received a few days ago your letter of the 13th ult., which was forwarded from New Orleans. Its consequent delay in reaching me will in part account for my tardiness in replying.

"Whilst I fully appreciate your unwillingness to comply with my request, I cannot agree in forming with you the conclusion that I should remain silent upon this important matter of history. I have, with your consent granted at the time of our conversation in Mobile, already mentioned to a number of our common friends, the fact of General Johnston's intention to abandon Richmond, after his retreat from Yorktown, as expressed to Mr. McFarland; and one of our friends, after listening with great interest to my assertion, and being informed that I had requested of you a written statement of the fact, remarked: 'If Charles M. Conrad told you this, he will give it to you in writing.'

"Moreover, I am not only indirectly but, I may say, directly concerned in this matter, from the fact that I have publicly stated that General Johnston foreshadowed to his corps commanders at New Hope Church, his intention to retreat to Macon, Georgia, during his campaign of '64 from Dalton. I am the only living witness of this historic truth; therefore, Mr. McFarland's testimony, through one of your prominence and character, becomes of great relative value to me.

"When I again have the leisure to continue my reply to the many unwarranted statements contained in General Johnston's book, I may find it necessary to bring forth also this important truth.

"I am yours truly,

"J. B. HOOD."

When I recall the different events with which the military career of General Johnston is connected, it is difficult to believe that he ever had any other fixed plan than that of retreat. Possibly the following paragraph in reference to a light engagement of General Hardee, on the 15th of March, 1865, near Averysboro', North Carolina, may indicate the nature of his expectations, after a surrender of Richmond, Atlanta, etc., etc., and a final retreat to the seashore, the last point of resistance: "That report, if correct, proves that the soldiers of General Sherman's Army had been demoralized by their course of life on the Southern plantations. Those soldiers, when fighting between Dalton and Atlanta, could not have been driven back repeatedly by a fourth of their number, with a loss so utterly insignificant." Was it General Johnston's policy to retreat till he had demoralized the enemy, and demoralized them by their course of life on Southern plantations? An easy victory even at such cost would, indeed, have been dearly bought.

I do not wish to be understood as, in any manner, questioning the courage of this General. He would have led men into action as gallantly as any soldier. But leading men into action is one thing, and ordering an Army into battle is another. To issue an order of great moment and simply to obey instructions, involve such different measures of responsibility that a distinct degree of moral courage is requisite to fulfil either duty. General Johnston has defended himself by charging me with recklessness, and exposure of my troops to "useless butchery." I may, therefore, be pardoned if I point out what I consider his main defects—the reason, in fact, why his name is not coupled with a single glorious victory in the annals of our four years' struggle, since it is, most assuredly, not because of lack of personal courage. It becomes necessary to express myself somewhat explicitly, in order that no misapprehension be engendered.

Caution and boldness are the two predominant qualities which characterize all soldiers of merit—I mean the caution and boldness tempered by wisdom, which such men as Napoleon I., Lee, Stonewall Jackson, Von Moltke, and Sir Garnet Wolseley have exhibited in so high a degree. These soldiers have shown themselves gifted with that intuition of the true warrior which rendered them bold in strategy, rapid in movement, and determined in battle. Observation has taught me that a commander may acquire sufficient caution by receiving hard blows, but he cannot acquire boldness. It is a gift from Heaven. A soldier whose quality of caution far exceeds that of boldness, can never be eminent in war. He cannot overcome nature, and experience that self-confidence requisite to order an Army into battle. If from pride or wounded vanity he make the venture, after long awaiting a more and more favorable opportunity, he will, as a rule, strike at the most unpropitious moment. Herein lies the deficiency of General Johnston. He is a man of courage and ability, and a fine organizer of an Army for the field; but he lacks the bold genius of Lee, and, consequently, will rarely, if ever, see sufficient chances in his favor—especially at the right time—to induce him to risk battle. Seven Pines is, I think, the only battle he attempted to inaugurate during the war, although it may be said that he commanded more men than any other Confederate officer. In this instance he had received information that a small body of the enemy had crossed the Chickahominy; he attempted to crush it with his entire force, and, even then, failed. He invariably throws up entrenchments, fortifies his line, and there remains in deliberation upon the best means to defeat the enemy without risking a general engagement, when, suddenly, he finds himself outflanked, and issues the usual order for retreat.

The same defect—want of decision and self-reliance—exhibits itself eminently in the lower grades of rank in an army. Among ten brave division commanders who, under orders, would lead their troops anywhere and everywhere, seldom will be found one who, in a position to act upon his own responsibility, will attack at a favorable moment, especially when detached and ordered to the rear of the enemy. If the inquiry be made of any enterprising, self-reliant division commander who has participated in many engagements, as to how often his request has been granted when, heavily engaged with superior numbers, he has called upon his neighboring division commander (whose troops were lying inactive) to furnish him the assistance of a brigade, he will reply that rarely has succor been afforded, even in the most critical moment of battle. His neighbor, albeit a man who knows not fear, was generally unwilling to act and give the necessary support without orders from his superior officer, because of his over-development of caution and his deficiency in boldness—the counterbalancing quality.

Again, few men are endowed with the capacity to execute such moves as those of Stonewall Jackson, at Second Manassas, and at Chancellorsville, for the reason that, whilst en route to the rear of the enemy, the appearance of a light squad of their cavalry will cause a majority of officers to halt, form line, reconnoitre, and thus lose time and the opportunity. Jackson's wagon train was attacked by Federal cavalry whilst he was marching to the rear and flank of Hooker, at Chancellorsville; he wisely paid little attention thereto, and moved boldly on towards the main object, and achieved a signal victory.

I shall allow to pass unnoticed, in this reply several statements of General Johnston which, although equally erroneous and illiberal in spirit, are too trivial to demand my attention. I shall, therefore, end this unpleasant discussion with a brief reference to his unpardonable conduct towards me, after he again assumed command in North Carolina. He was not unmindful that he had again been restored to power. This new acquisition of authority, he determined should be felt by those who had ventured to oppose his policy, and contradict his statements. Accordingly, as I was en route for the Trans-Mississippi Department, under orders to bring to the support of General Lee all the troops that would follow me, I received, at Chester, South Carolina, the following telegram:

"SMITHFIELD, *April 4th, 1865.*

"LIEUTENANT GENERAL J. B. HOOD.

"After reading your report, as submitted, I informed General Cooper, by telegraph, that I should prefer charges against you as soon as I have leisure to do so, and desired him to give you the information.

"J. E. JOHNSTON."

I replied as follows:

"CHESTER, SOUTH CAROLINA, *April 4th, 1865.*

"GENERAL J. E. JOHNSTON, *Smithfield, N. C.*

"Your telegram of this date received, informing me that you intended, as soon as you had leisure, to prefer charges against me. I am under orders for the Trans-Mississippi Department. I shall inquire of General Cooper whether I am to await my trial or proceed as ordered. I will be ready to meet any charges you may prefer.

"J. B. HOOD."

On the following day I applied to the War Department for a Court of Inquiry.

"CHESTER, SOUTH CAROLINA, *April 5th, 1865.*

"GENERAL S. COOPER.

"I have the honor to request that a Court of Inquiry be assembled at the earliest practicable moment to investigate and report upon the facts and statements contained in my official report of the operations of the Army of the Tennessee.

"J. B. HOOD,
"Lieutenant General."

I received the following in reply:

"DANVILLE, *April 5th, 1865.*

"LIEUTENANT GENERAL J. B. HOOD.

"Proceed to Texas as heretofore ordered.

"S. COOPER, A. I. G."

"DANVILLE, *April 7th, 1865.*

"LIEUTENANT GENERAL J. B. HOOD.

"A Court of Inquiry cannot be convened in your case at present. You will proceed to Texas as heretofore ordered.

"S. COOPER, A. I. G."

Had I been granted a Court of Inquiry at that date, I would have produced stronger testimony than I have given, even at this late period, in relation to the points in controversy between General Johnston and myself.

This attempt to summons me before a Court Martial was his final effort, during the war, to asperse the character of a brother officer who had always been true to duty, but whose unpardonable crime was having been appointed to supersede him in the command of the Army of Tennessee.

CHAPTER X.

SIEGE OF ATLANTA—DIFFICULTIES OF THE SITUATION—BATTLE OF THE 20TH OF JULY.

NOTWITHSTANDING the manifold difficulties and trials which beset me at the period I was ordered to relieve General Johnston, and which, because of unbroken silence on my part, have been the occasion of much injustice manifested in my regard, I formed no intention, till the appearance of General Sherman's Memoirs, to enter fully into the details of the siege of Atlanta, the campaign to the Alabama line, and that which followed into Tennessee.

A feeling of reluctance to cause heart-burnings within the breast of any Confederate, who fulfilled his duty to the best of his ability, has, hitherto, deterred me from speaking forth the truth. Since, however, military movements with which my name is closely connected, have been freely and publicly discussed by different authors, whose representations have not always been accurate, I feel compelled to give an account of the operations of the Army of Tennessee, whilst under my direction.

As already mentioned, the order, assigning me to the command of that Army, was received about 11 p. m., on the 17th of July. My predecessor, unwilling to await even the dawn of day, issued his farewell order that memorable night. In despite of my repeated and urgent appeals to him to pocket all despatches from Richmond, to leave me in command of my own corps, and to fight the battle for Atlanta, he deserted me the ensuing afternoon. He deserted me in violation of his promise to remain and afford me the advantage of his counsel, whilst I shouldered all responsibility of the contest.

I reiterate that it is difficult to imagine a commander placed at the head of an Army under more embarrassing circumstances than those against which I was left to contend on the evening of the 18th of July, 1864. I was, comparatively, a stranger to the Army of Tennessee. Moreover General Johnston's mode of warfare formed so strong a contrast to the tactics and strategy which were practiced in Virginia, where far more satisfactory results were obtained than in the West, that I have become a still more ardent advocate of the Lee and Jackson school. The troops of the Army of Tennessee had for such length of time been subjected to the ruinous policy pursued from Dalton to Atlanta that they were unfitted for united action in pitched battle. They had, in other words, been so long habituated to security behind breastworks that they had become wedded to the "timid defensive" policy, and naturally regarded with distrust a commander likely to initiate offensive operations.

The senior Corps Commander considered he had been supplanted through my promotion, and thereupon determined to resign, in consequence, I have no doubt, of my application to President Davis to postpone the order transferring to me the command of the Army; he however, altered his decision, and concluded to remain with his corps.

The evening of the 18th of July found General Johnston comfortably quartered at Macon, whilst McPherson's and Schofield's Corps were tearing up the Georgia Railroad, between Stone Mountain and Decatur; Thomas's Army was hastening preparations to cross Peach Tree creek, within about six miles of Atlanta; and I was busily engaged in hunting up the positions of, and establishing communication with Stewart's and Hardee's Corps, since I did not know where they were posted, when General Johnston disappeared so unexpectedly and left me in this critical position.

Not till I read Sherman's Memoirs, was I aware of McPherson's so close proximity to Atlanta at an early hour on the 18th of July. In truth, a few enterprising scouts thrown out that afternoon from his columns, in the direction of the Macon Railroad might have captured my predecessor on his retreat to Macon.

Sherman says (vol. II, pages 71, 72):

> "On the 18th all the Armies moved on a general right wheel, Thomas to Buckhead, forming line of battle facing Peach Tree creek; Schofield was on his left, and McPherson well on towards the railroad between Stone Mountain and Decatur, which he reached at 2 p. m. of that day, about four miles from Stone Mountain, and seven miles east of Decatur, and there he turned toward Atlanta, breaking up the railroad as he progressed, his advance guard reaching Decatur about night, where he came into communication with Schofield's troops, which had also reached Decatur."

It thus appears that on the afternoon of the 18th the enemy was in Decatur, almost at the gates of Atlanta. This intelligence must have been communicated to General Johnston by the cavalry, after he left me to ride into the city with the promise to return toward evening, as he was virtually Commander-in-Chief up to the moment of his sudden departure. I had consumed a great portion of that day in vain endeavors to adjust the difficulties in the way of his retention in command, by earnest representations to him, on one hand, and, on the other, by telegraphing to Richmond in the hope of accomplishing this object. Although he had published his farewell order the night previous, I had not, owing to the foregoing reasons, assumed command. He had agreed to issue orders in my name, and, in reality, I did not become Commander-in-Chief until about night of that day, when I received information of his departure.

Much confusion necessarily arose during this interval; and this condition of affairs accounts for the circumstance, which must seem strange to military men, that at this late date I am apprised for the first time, and through Sherman's Memoirs, of the presence of the enemy's left wing, at 2 p. m. on the 18th of July, upon the railroad leading to Augusta. It must seem equally strange that, if I was regarded as chief in command, this important movement was not made known to me at headquarters by our cavalry, which was, generally, very prompt in reporting all such information. I cannot but think, therefore, that General Johnston was cognizant before 4 o'clock that day, and before his departure for Macon, of the enemy's presence on the Augusta Railroad, within six or eight miles of Atlanta. If such is not the case, our cavalry, stationed upon the right, neglected most unpardonably its duty—which supposition I am not inclined to admit.

The statement in my official report* that McPherson was at Decatur on the morning of the 19th, is proof of my ignorance of the circumstance on the 18th.

These facts give evidence of the trying position in which I was placed at this juncture, and this last move of the enemy may somewhat account for the inexplicable conduct and disappearance of General Johnston who, at this critical moment, was unwilling to share with me the responsibility of the issue.

I will now turn from the many unpleasant occurrences interwoven in the history of that day, and endeavor to show in what manner General Sherman exposed, on his approach to Atlanta, the Federal Army to successful attack by our troops; and, at the same time, state why, in my opinion, after our discovery of his blunders, the Confederate Army did not succeed in defeating and routing his forces.

These premises may seem bold, especially since defeat was our fortune, and victory the boast of our adversary.

After having established communication with the corps and

---

* Appendix, p. 320.

the cavalry of the Army during the forepart of the night, I found myself, upon the morning of the 19th, in readiness to fulfil these grave duties devolving upon me.

Our troops had awakened in me heartfelt sympathy, as I had followed their military career with deep interest from early in May of that year. I had witnessed their splendid condition at that period; had welcomed with pride the fine body of reinforcements under General Polk; but, with disappointment, I had seen them, day after day, turn their back upon the enemy, and lastly cross the Chattahoochee river on the night of the 9th of July with one-third of their number lost—the men downcast, dispirited, and demoralized. Stragglers and deserters, the captured and the killed, could not now, however, be replaced by recruits, because all the recruiting depots had been drained to reinforce either Lee or Johnston. I could, therefore, but make the best dispositions in my power with the reduced numbers of the Army, which opposed a force of one hundred and six thousand (106,000) Federals, buoyant with success and hope, and who were fully equal to one hundred and forty thousand (140,000) such troops as confronted Johnston at Dalton, by reason of their victorious march of a hundred miles into the heart of the Confederacy.

Accordingly, on the night of the 18th and morning of the 19th, I formed line of battle facing Peach Tree creek; the left rested near Pace's Ferry road, and the right covered Atlanta. I was informed on the 19th that Thomas was building bridges across Peach Tree creek; that McPherson and Schofield were well over toward, and even on, the Georgia Railroad, near Decatur. I perceived at once that the Federal commander had committed a serious blunder in separating his corps, or Armies by such distance as to allow me to concentrate the main body of our Army upon his right wing, whilst his left was so far removed as to be incapable of rendering timely assistance. General Sherman's violation of the established maxim that an Army should always be held well within hand, or its detachments within easy supporting distance, afforded one of the most favorable occasions for complete victory which could have been offered; especially as it presented an opportunity, after crushing his right wing, to throw our entire force upon his left. In fact, such a blunder affords a small Army the best, if not the sole, chance of success when contending with a vastly superior force.

Line of battle having been formed, Stewart's Corps was in position on the left, Hardee's in the centre, and Cheatham's on the right. Orders were given to Generals Hardee and Stewart to observe closely and report promptly the progress of Thomas in the construction of bridges across Peach Tree creek and the passage of troops. General Cheatham was directed to reconnoitre in front of his left; to erect, upon that part of his line, batteries so disposed as to command the entire space between his left and Peach Tree creek, in order to completely isolate McPherson and Schofield's forces from those of Thomas; and, finally, to thoroughly entrench his line. This object accomplished, and Thomas having partially crossed the creek and made a lodgment on the east side within the pocket formed by Peach Tree creek and the Chattahoochee river, I determined to attack him with two corps—Hardee's and Stewart's, which constituted the main body of the Confederate Army—and thus, if possible, crush Sherman's right wing, as we drove it into the narrow space between the creek and the river.

Major General G. W. Smith's Georgia State troops were posted on the right of Cheatham, and it was impossible for Schofield or McPherson to assist Thomas without recrossing Peach Tree creek in the vicinity of Decatur, and making on the west side a detour which necessitated a march of not less than ten or twelve miles, in order to reach Thomas's bridges across this creek. I immediately assembled the three corps commanders, Hardee, Stewart, and Cheatham, together with Major General G. W. Smith, commanding Georgia State troops, for the purpose of giving orders for battle on the following day, the 20th of July.

Map 3

BATTLES AROUND ATLANTA.
Scale: 1 inch = 4 miles.
N.
CHATTAHOOCHEE RIVER
Nickajack Cr.
Sweetwater Cr.
Racers Fy.
Howells Fy.
Buckhead
Peach Tree Cr.
Howells M.
Peach Tree Ch.
Mason's M.
Hooper M.
Parts M.
Proctors Cr.
Western & Atlantic R.R.
White House
Decatur
Green or Howells Fy.
Lick Skillet
Ellotts M.
Howards Still
Bakers Fy.
Buzzard Roost
ATLANTA
White Hall
Donohue M.
Dollans M.
Utoy
Utoy Cr.
Moores M.
Alexanders M.
Sandtown
Herring Ch.
Intrenchment
Dry Pond
Connollys M.
Cobbs M.
Panthersville
Germans Fy.
Owl Ch.
Willis M.
Terry's M.
Hopewell Ch.
South River
Fulton Copper Mine
Camp Cr.
McGilead Ch.
East Point
Poolville
Woodville
Liberty Ch.
Cones Ch.
Hornsby M.
Minns
Camp Cr. Ch.
Biggs
Birds Ch.
Campbellton
Sherman after Demonstration
Trimbles M.
Rough & Ready
Deep Cr.
Smyrna Ch.
Huis M.
Red Oak
Quick St.
Eahes Ch.
K&L

I here quote from my official report written soon after these events:

"On the morning of the 19th, the dispositions of the enemy were substantially as follows: 'The Army of the Cumberland, under Thomas, was in the act of crossing Peach Tree creek. This creek, forming a considerable obstacle to the passage of an army, runs in a northeasterly direction, emptying into the Chattahoochee river near the railroad crossing. The Army of the Ohio, under Schofield, was also about to cross east of the Buckhead road. The Army of the Tennessee, under McPherson, was moving on the Georgia Railroad at Decatur. Finding it impossible to hold Atlanta without giving battle, I determined to strike the enemy while attempting to cross this stream. My troops were disposed as follows: Stewart's Corps on the left, Hardee's in the centre, and Cheatham's on the right entrenched. My object was to crush Thomas's Army before he could fortify himself, and then turn upon Schofield and McPherson. To do this, Cheatham was ordered to hold his left on the creek, in order to separate Thomas's Army from the forces on his (Thomas's) left. Thus I should be able to throw two corps, Stewart's and Hardee's, against Thomas. Specific orders were carefully given these Generals, in the presence of each other, as follows: The attack was to begin at 1 p. m.; the movement to be by division, *en echelon* from the right, at a distance of about one hundred and fifty yards; the effort to be to drive the enemy back to the creek, and then towards the river, into the narrow space formed by the river and creek; everything on our side of the creek to be taken at all hazards, and to follow up as our success might permit. Each of these Generals was to hold a division in reserve. Owing to the demonstrations of the enemy on the right, it became necessary to extend Cheatham a division front to the right. To do this, Hardee and Stewart were each ordered to extend a half division front to close the interval. Foreseeing that some confusion and delay might result, I was careful to call General Hardee's attention to the importance of having a staff officer on his left to see that the left did not take more than a-half division front. This, unfortunately, was not attended to, and the line closed to the right, causing Stewart to move two or three times the proper distance. In consequence of this, the attack was delayed until nearly 4 p. m.'"

The three corps commanders, together with General G. W. Smith, were assembled not only for the purpose of issuing to them orders for battle, but with the special design to deliver most explicit instructions in regard to their respective duties.

I sought to "make assurance doubly sure" by direct interrogatory; each was asked whether or not he understood his orders. All replied in the affirmative. I was very careful in this respect, inasmuch as I had learned from long experience that no measure is more important, upon the eve of battle, than to make certain, in the presence of the commanders, that each thoroughly comprehends his orders. The usual discretion allowed these officers in no manner diminishes the importance of this precaution.

I also deemed it of equal moment that each should fully appreciate the imperativeness of the orders then issued, by reason of the certainty that our troops would encounter hastily constructed works thrown up by the Federal troops, which had been foremost to cross Peach Tree creek. Although a portion of the enemy would undoubtedly be found under cover of temporary breastworks, it was equally certain a larger portion would be caught in the act of throwing up such works, and just in that state of confusion to enable our forces to rout them by a bold and persistent attack. With these convictions I timed the assault at 1 p. m., so as to surprise the enemy in their unsettled condition.

As stated in my official report, the charge was unfortunately not made till about 4 o'clock p. m., on account of General Hardee's failure to obey my specific instructions in regard to the extension of the one-half division front to the right, in order to afford General Cheatham an advantageous position to hold in check McPherson and Schofield. The result was not, however, materially affected by this delay, since the Federals were completely taken by surprise.

General Stewart carried out his instructions to the letter; he moreover appealed in person to his troops before going into action, and informed them that orders were imperative they should carry everything, at all hazards, on their side of Peach Tree creek; he impressed upon them that they should not halt before temporary breastworks, but charge gallantly over every obstacle and rout the enemy. It was evident that after long-continued use of entrenchments, General Stewart deemed a personal appeal to his soldiers expedient. An address from a corps commander to his troops, upon the eve of battle, is always productive either of great good or evil, according to the spirit in which it is spoken. For this reason, commanders of large bodies should exercise extreme caution in the expression of their sentiments, even in the presence of staff officers. Every word, portending probable results, passes like an electric spark through the entire command. It is, therefore, in the power of an officer to inspirit his men, and incite them to deeds of valor in the hour of battle, as well as to depress and demoralize them by an expression of despondency, one word foreshadowing the possibility of defeat.

General Stewart and his troops nobly performed their duty in the engagement of the 20th. At the time of the attack, his corps moved boldly forward, drove the enemy from his works, and held possession of them until driven out by an enfilade fire of batteries placed in position by General Thomas.

The following extracts from the reports of Generals Stewart and Featherston, touching the battle of the 20th, will be read with interest:

"The plan was for the divisions (commencing on Hardee's right) to move forward, successively, *en echelon*, at intervals of some two hundred yards, to attack the enemy, drive him back to the creek, and then press down the creek to the left. Should the enemy be found entrenched, his works were to be carried; everything on our side of the creek was to be taken, and our crossing to the other side of the creek was to depend on our success. Such were the instructions of the General Commanding to General Hardee and myself. I was to hold a division in reserve. It seems a division had been withdrawn from the lines on the right of Hardee's Corps. His corps and mine were to close to the right far enough to cover the space vacated by this division—the space to be divided between the two corps. This would have shifted my line a-half division front to the right—perhaps at most—half-a-mile. At 1 o'clock I found the left of Hardee's Corps just beginning to shift to the right. Feeling that this change was not important, and that not a moment was to be lost in making the attack contemplated, a staff officer was despatched to the Commanding General to inform him of the fact, and requesting an order to stop the movement to the right and commence the forward movement. The result was, however, that to keep up connection with the other corps, my line was moved a mile and a-half or two miles to the right. * * * My instructions to division commanders, and through them to brigade and regimental commanders, were to move forward and attack the enemy; if found entrenched, to fix bayonets and carry his works, to drive him back to the creek, and then press down the creek; that we were to carry everything in our front on our side of the creek. * * * These commanders, their officers and men, behaved entirely to my satisfaction, and I cannot but think, had the plan of the battle, as I understood it, been carried out fully, we would have achieved a great success."

General Featherston in his official report writes:

"The plan of the battle, as explained to me, was as follows: The attack was to begin on the extreme right of the Army. General Hood's old corps and General Hardee's were both on my right. The troops were to advance *en echelon* by divisions, beginning on the extreme right; the first division advancing some three hundred yards to the front before the second moved. The same order was to be observed down the entire line from right to left, extending through all three of the army corps. Each division when it reached Peach Tree creek was to oblique to the left, and sweep down the creek, and thereby make the attack upon the enemy, one upon his front and left flank at the same time. My orders were to fix bayonets and charge their works when we reached them, to stop for no obstacle, however formidable, but to make the attack a desperate one. I was informed that the same orders had been delivered by the Commander-in-Chief, General Hood, to each and every army corps. I thought the battle had been well planned, and heard it spoken of by my associates in arms in terms of commendation. The whole corps, so far as I heard an expression of their opinions, anticipated a brilliant victory.

"I was struck with surprise, at the time we moved to the front, that no guns, either artillery or small arms, were heard on our right, save a feeble skirmish. I supposed from hearing no firing on our right, and knowing that many divisions had had time to reach the creek, that they had found no enemy in their front. Had the attack been vigorously made by all the troops on our right, and the plan of the battle been strictly carried out, I then believe, and still believe, the victory would have been a brilliant one, and the Federal forces on the south side of Peach Tree creek would have been all either killed, wounded or captured. The orders seem to have been misunderstood by the troops on our right, or for some cause not fully carried out."

Unfortunately, the corps on Stewart's right, although composed of the best troops in the Army, virtually accomplished nothing. In lieu of moving the half division front promptly to the right, attacking as ordered, and supporting Stewart's gallant assault, the troops of Hardee—as their losses on that day indicate—did nothing more than skirmish with the enemy. Instead of charging down upon the foe as Sherman represents Stewart's men to have done, many of the troops, when they discovered that they had come into contact with breastworks, lay down and, consequently, this attempt at pitched battle proved abortive.

I was at the time unable to discover a satisfactory reason for which an united attack by two corps d'armee, at even 4 o'clock in the day, should have failed to destroy Thomas's Army, which was protected by only slight entrenchments and was situated within a pocket formed by two streams difficult of passage. I was deeply concerned and perplexed, as I sought to divine the cause of misfortune—especially of failure on the part of Hardee's Corps to attack according to explicit instructions.

General Sherman writes as follows, in regard to this engagement:*

"On the 19th the three Armies were converging towards Atlanta, meeting such feeble resistance that I really thought the enemy intended to evacuate the place. McPherson was moving astride of the railroad, near Decatur; Schofield along a road leading toward Atlanta, by Colonel Howard's house and the distillery; and Thomas was crossing 'Peach Tree' in line of battle, building bridges for nearly every division as deployed. There was quite a gap between Thomas and Schofield, *which I endeavored to close by drawing two of Howard's Divisions nearer Schofield.* On the 20th I was with General Schofield near the centre, and soon after noon heard heavy firing in front of Thomas's right, which lasted an hour or so, and then ceased. I soon learned that the enemy had made a furious sally, the blow falling on Hooker's Corps (the Twentieth), and partially on Johnston's Division of the Fourteenth, and Newton's of the Fourth. The troops had crossed Peach Tree creek,

*Sherman's Memoirs, vol. II, pages 72, 73.

were deployed, but at the time were resting for noon, when, without notice, the enemy came pouring out of their trenches down upon them, they became commingled, and fought in many places hand to hand. General Thomas happened to be near the rear of Newton's Division, and got some field batteries in good position, on the north side of Peach Tree creek, from which he directed a furious fire upon a mass of the enemy, which was passing round Newton's left and exposed flank. After a couple of hours of hard and close conflict, the enemy retired slowly within his trenches, leaving his dead and many wounded on the field."

CHAPTER XI.

SIEGE OF ATLANTA—BATTLE 22D OF JULY—HARDEE—GENERAL FRANK BLAIR'S LETTER.

THE failure on the 20th, rendered urgent the most active measures, in order to save Atlanta even for a short period. Through the vigilance of General Wheeler, I received information, during the night of the 20th, of the exposed position of McPherson's left flank; it was standing out in air, near the Georgia Railroad between Decatur and Atlanta, and a large number of the enemy's wagons had been parked in and around Decatur. The roads were in good condition, and ran in the direction to enable a large body of our Army to march, under cover of darkness, around this exposed flank, and attack in rear.

I determined to make all necessary preparations for a renewed assault; to attack the extreme left of the Federals in rear and flank, and endeavor to bring the entire Confederate Army into united action.

Accordingly, Hardee's and Stewart's Corps resumed their former positions. Colonel Prestman, chief engineer, was instructed to examine at once the partially completed line of works toward Peach Tree creek, which General Johnston had ordered to be constructed for the defence of Atlanta, and to report, at the earliest moment, in regard to their fitness to be occupied by Stewart's and Cheatham's Corps, together with the Georgia State troops, under General G. W. Smith. The report was received early on the morning of the 21st, to the effect that the line established by Johnston, was not only too close to the city and located upon too low ground, but was totally inadequate for the purpose designed; that Sherman's line, which extended from the vicinity of Decatur almost to the Dalton Railroad, north of Atlanta, rendered necessary the construction of an entirely new line, and upon more elevated ground.

The chief engineer was thereupon directed to prepare and stake off a new line, and to employ his entire force, in order that the troops might occupy the works soon after dark on the night of the 21st, and have time to aid in strengthening their position before dawn of next morning. This task was soon executed through the skill and energy of Colonel Prestman and his assistants. Generals Stewart, Cheatham, and G. W. Smith, were instructed to order their division and brigade commanders to examine before dark the ground to be occupied by their respective troops, so as to avoid confusion, or delay, at the time of the movement.

General Hardee, who commanded the largest corps, and whose troops were comparatively fresh, as they had taken but little part in the attack of the previous day, was ordered to hold his forces in readiness to move promptly at dark that night—the 21st. I selected Hardee for this duty, because Cheatham had, at that time, but little experience as a corps commander, and Stewart had been heavily engaged the day previous.

The position of the enemy during the 21st remained, I may say, unchanged, with the exception that Schofield and McPherson had advanced slightly toward Atlanta. To transfer after dark our entire line from the immediate presence of the enemy to another line around Atlanta, and to throw Hardee, the same night, entirely to the rear and flank of McPherson—as Jackson was thrown, in a similar movement, at Chancellorsville and Second Manassas—and to initiate the offensive at daylight, required no small effort upon the part of the men and officers. I hoped, however, that the assault would result not only in a general battle, but in a signal victory to our arms.

It was absolutely necessary these operations should be executed that same night, since a delay of even twenty-four hours would allow the enemy time to further entrench, and afford Sherman a chance to rectify, in a measure, his strange blunder in separating Thomas so far from Schofield and McPherson. Sherman evidently perceived his error, as the following extract from his Memoirs denotes: *

"There was quite a gap between Thomas and Schofield, which I endeavored to close by drawing two of Howard's Divisions near Schofield."

I well knew he would seek to retrieve his oversight at the earliest possible moment; therefore, I determined to forestall his attempt, and to make another effort to defeat the Federal Army. No time was to be lost in taking advantage of this second unexpected opportunity to achieve victory and relieve Atlanta.

I was convinced that McPherson and Schofield intended to destroy not only the Georgia Railroad, but likewise our main line of communication, the railroad to Macon. It is now evident the blow on the 20th checked the reckless manner of moving, which had so long been practiced by the enemy, without fear of molestation, during the Dalton–Atlanta campaign. The rap of warning received by Thomas, on Peach Tree creek, must have induced the Federal commander to alter his plan. He says in relation thereto: †

"During the night (21st), I had full reports from all parts of our line, most of which was partially entrenched as against a sally, and finding that McPherson was stretching out too much on his left flank, I wrote him a note early in the morning not to extend so much by his left; for

* Sherman's Memoirs, vol. II, page 72.

† Sherman's Memoirs, vol. II, page 74.

we had not troops enough to completely invest the place, and I intended to destroy utterly all parts of the Augusta Railroad to the east of Atlanta, then to withdraw from the left flank and add to the right."

Thus was situated the Federal Army at the close of night, on the 21st: it was but partially entrenched; Schofield and McPherson were still separated from Thomas, and at such distance as to compel them to make a detour of about twelve miles, in order to reach the latter in time of need.

The Confederate Army occupied the same position, at dark, as prior to the attack of the 20th. The new line around the city, however, had been chosen; each corps commander fully advised of the ground assigned to him, and the special duty devolving upon him; working parties had been detailed in advance from the corps of Stewart and Cheatham, and from the Georgia State troops; rations and ammunition had been issued, and Hardee's Corps instructed to be in readiness to move at a moment's warning.

The demonstrations of the enemy upon our right, and which threatened to destroy the Macon Railroad—our main line for receiving supplies—rendered it imperative that I should check, immediately, his operations in that direction; otherwise Atlanta was doomed to fall at a very early day. Although the attack of the 20th had caused Sherman to pause and reflect, I do not think he would have desisted extending his left toward our main line of communication, had not the events occurred which I am about to narrate.

As already stated, every preparation had been carefully made during the day of the 21st. I had summoned, moreover, to my headquarters the three corps commanders, Hardee, Stewart, and Cheatham, together with Major General Wheeler, commanding cavalry corps, and Major General G. W. Smith, commanding Georgia State troops. The following minute instructions were given in the presence of all assembled, in order that each might understand not only his own duty, but likewise that of his brother corps commanders; by this means I hoped each officer would know what support to expect from his neighbor, in the hour of battle.

Stewart, Cheatham, and G. W. Smith, were ordered to occupy soon after dark the positions assigned them in the new line round the city, and to entrench as thoroughly as possible. General Shoupe, chief of artillery, was ordered to mass artillery on our right. General Hardee was directed to put his corps in motion soon after dusk; to move south on the McDonough road, across Entrenchment creek at Cobb's Mills, and to *completely* turn the left of McPherson's Army and attack at daylight, or as soon thereafter as possible. He was furnished guides from Wheeler's cavalry, who were familiar with the various roads in that direction; was given clear and positive orders to detach his corps, to swing away from the main body of the Army, and to march entirely around and to the rear of McPherson's left flank, even if he was forced to go to or beyond Decatur, which is only about six miles from Atlanta.*

Major General Wheeler was ordered to move on Hardee's right with all the cavalry at his disposal, and to attack with Hardee at daylight. General Cheatham, who was in line of battle on the right and around the city, was instructed to take up the movement from his right as soon as Hardee succeeded in forcing back, or throwing into confusion, the Federal left, and to assist in driving the enemy down and back upon Peach Tree creek, from right to left. General G. W. Smith would, thereupon, join in the attack. General Stewart, posted on the left, was instructed not only to occupy and keep a strict watch upon Thomas, in order to prevent him from giving aid to Schofield and McPherson, but to engage the enemy the instant the movement became general, *i. e.*, as soon as Hardee and Cheatham succeeded in driving the Federals down Peach Tree creek and near his right. Though the movement assigned General Hardee, on this occasion, was a very simple one, it is,

---

* Hood's Official Report, Appendix p. 321.

as I have remarked in my reply to General Johnston, rare to find one out of ten brave division commanders, who is capable of swinging away from the main army and attacking in rear as Jackson did at Second Manassas and Chancellorsville. The march, however, in this instance, was so short—Decatur being only six miles from Atlanta, and our cavalry had so often passed back and forth over the roads, day and night, in bringing dispatches from Wheeler to Army headquarters, and consequently were so thoroughly familiar with the different routes, that I considered Hardee's move one merely within the lines of our cavalry; that no special quality, such as Jackson possessed, was required; that he had simply to follow the guides furnished him to Decatur, and attack as ordered.

Thus orders were given to attack from right to left, and to press the Federal Army down and against the deep and muddy stream in their rear. These orders were carefully explained again and again, till each officer present gave assurance that he fully comprehended his duties. The following extract will disclose the situation and surmises of the enemy on the morning of the battle:*

"In the morning (22d) we found the strong line of parapet, 'Peach Tree line,' to the front of Schofield and Thomas, abandoned, and our lines were advanced rapidly close up to Atlanta. For some moments I supposed the enemy intended to evacuate, and in person was on horseback at the head of Schofield's troops, who had advanced in front of the Howard House to some open ground, from which we could plainly see the whole rebel line of parapets, and I saw their men dragging up from the intervening valley, by the distillery, trees and saplings for abatis. Our skirmishers found the enemy down in this valley, and we could see the rebel main line strongly manned, with guns in position at intervals. Schofield was dressing forward his lines, and I could hear Thomas further to the right engaged, when General McPherson and his staff rode up. We went back to the Howard House, a double frame building with a porch, and sat on the steps, discussing the chances of battle, and of Hood's general character. McPherson had also been of the same class at West Point with Hood, Schofield, and Sheridan. We agreed that we

* Sherman's Memoirs, vol. II, pages 74, 75.

ought to be unusually cautious and prepared at all times for sallies and for hard fighting, because Hood, though not deemed much of a scholar, or of great mental capacity, was undoubtedly a brave, determined, and rash man; and the change of commanders at that particular crisis argued the displeasure of the Confederate Government with the cautious but prudent conduct of General Joe Johnston."

At dawn on the morning of the 22d Cheatham, Stewart, and G. W. Smith, had, by alternating working parties during the night previous, not only strongly fortified their respective positions, but had kept their men comparatively fresh for action, and were in readiness to act as soon as the battle was initiated by Hardee who was supposed to be at that moment in rear of the adversary's flank.

I took my position at daybreak near Cheatham's right, whence I could observe the left of the enemy's entrenchments which seemed to be thrown back a short distance on their extreme left. After awaiting nearly the entire morning, I heard, about ten or eleven o'clock, skirmishing going on directly opposite the left of the enemy, which was in front of Cheatham's right and Shoupe's artillery. A considerable time had elapsed when I discovered, with astonishment and bitter disappointment, a line of battle composed of one of Hardee's divisions advancing directly against the entrenched flank of the enemy. I at once perceived that Hardee had not only failed to turn McPherson's left, according to positive orders, but had thrown his men against the enemy's breastworks, thereby occasioning unnecessary loss to us, and rendering doubtful the great result desired. In lieu of completely turning the Federal left and taking the entrenched line of the enemy in reverse, he attacked the retired wing of their flank, having his own left almost within gunshot of our main line around the city. I then began to fear that his disregard of the fixed rule in war that one danger in rear is more to be feared than ten in front—in other words, that one thousand men in rear are equal to ten thousand in front—would cause us much embarrassment, and place his corps at great disadvantage, notwithstanding he had held success within easy grasp. It had rested in his power to rout McPherson's Army by simply moving a little further to the right, and attacking in rear and flank instead of assaulting an entrenched flank. I hoped, nevertheless, this blunder would be remedied, at least, in part, by the extreme right of his line lapping round, during the attack, to the rear of McPherson.

I anxiously awaited tidings from the scene of action while listening attentively to what seemed a spirited engagement upon that part of the field. The following extract testifies to the uneasiness which Sherman experienced at the possibility of an attack upon his rear and flank:*

"Although the sound of musketry grew in volume, I was not so much disturbed by it as by the sound of artillery back toward Decatur."

This alarming sound proceeded from the guns of the gallant Wheeler, in the direction of Decatur, whence I hoped, momentarily, to hear a continuous roar of musketry, accompanied by the genuine Confederate shout from Hardee's entire Corps, as it advanced and drove the enemy down Peach Tree creek between our general line of battle and that formidable stream. Although the troops of Hardee fought, seemingly, with determination and spirit, there were indications that the desired end was not being accomplished. The roar of musketry occurring only at intervals strengthened this impression, and a staff officer was dispatched to General Hardee to know the actual result.

During the early afternoon I received information that the attack had been, in part, successful, but had been checked in consequence of our troops coming in contact with different lines of entrenchments, several of which they had carried and held. Fearing a concentration of the enemy upon Hardee, I commanded General Cheatham, about 3 p. m., to move forward with his corps, and attack the position in his front, so as

---

* Sherman's Memoirs, vol. II, page 76.

to, at least, create a division. The order was promptly and well executed, and our troops succeeded in taking possession of the enemy's defences in that part of the field. A heavy enfilade fire, however, forced Cheatham to abandon the works he had captured.

Major General G. W. Smith, perceiving that Cheatham had moved out on his left, and having thoroughly comprehended all the orders relative to the battle, moved gallantly forward with his State troops in support of Cheatham's attack, but was eventually forced to retire on account of superiority of numbers in his front. The militia, under his leadership, acted with distinction on this occasion, and Georgia has reason to congratulate herself that her troops were under the command of a soldier of the ability and skill of General G. W. Smith.

Hardee bore off as trophies eight guns and thirteen stands of colors, and, having rectified his line, remained in the presence of the enemy. Cheatham captured five guns and five or six stands of colors.

Notwithstanding the non-fulfilment of the brilliant result anticipated, the partial success of that day was productive of much benefit to the Army. It greatly improved the *morale* of the troops, infused new life and fresh hopes, arrested desertions, which had hitherto been numerous, defeated the movement of McPherson and Schofield upon our communications, in that direction, and demonstrated to the foe our determination to abandon no more territory without, at least, a manful effort to retain it.

I cannot refrain from mentioning the noble and gallant old hero, Major General W. H. S. Walker, who fell at the head of his division whilst bravely leading it into battle on the 22d of July. He was an officer of the old Army, had served with great distinction in the Mexican war, and was generally beloved by officers and men. On the night of the 21st, shortly before joining in Hardee's line of march with his troops, he rode by my headquarters, called me aside, and, with characteristic frankness, expressed his appreciation of the grave responsibilities attached to the position in which I had been placed; assured me that he full well understood the condition of the Army, after our protracted retreat from Dalton, and wished me to know, before he entered into battle, that he was with me in heart and purpose, and intended to abide by me through all emergencies. During the early afternoon of the ensuing day, I received the painful intelligence of his death; and I am certain that those officers and men who came within the sphere of his genial presence, will unite in the verdict that no truer or braver man ever fell upon the field of battle.

In connection with this sad event, I will record also the death of my classmate and friend in boyhood, General McPherson, which occurred the same day, and the announcement of which caused me sincere sorrow. Although in the same class, I was several years his junior, and, unlike him, was more wedded to boyish sports than to books. Often, when we were cadets, have I left barracks at night to participate in some merry-making, and early the following morning have had recourse to him to help me over the difficult portions of my studies for the day. Since we had graduated in June, 1853, and had each been ordered off on duty in different directions, it had not been our fortune to meet. Neither the lapse of years, nor the difference of sentiment which led us to range ourselves on opposite sides in the late war, had lessened my friendship; indeed the attachment, formed in early youth, was strengthened by my admiration and gratitude for his conduct toward our people in the vicinity of Vicksburg. His considerate and kind treatment of them stood in bright contrast to the course pursued by many Federal officers; and his acts were ever characterized by those gentlemanly qualities which distinguished him as a boy. No soldier fell in the enemy's ranks, whose loss caused me equal regret.

It became apparent almost immediately after the battle of the 22d that Sherman would make an attack upon our left, in order to destroy the Macon Railroad; and, from that moment, I may say, began the siege of Atlanta. The battles of the

20th and 22d checked the enemy's reckless manner of moving, and illustrated effectually to Sherman the danger of stretching out his line in such a manner as to form extensive gaps between his Corps, or Armies, as he admits he did at Rocky-face Ridge and New Hope Church, and, as I have no doubt, he did many times with impunity, when driving us before him through the mountain fastnesses of Georgia.

My failure on the 20th, and 22d, to bring about a general pitched battle arose from the unfortunate policy pursued from Dalton to Atlanta, and which had wrought such demoralization amid rank and file as to render the men unreliable in battle. I cannot give a more forcible, though homely, exemplification of the *morale* of the troops, at that period, than by comparing the Army to a team which has been allowed to balk at every hill: one portion will make strenuous efforts to advance, whilst the other will refuse to move and thus paralyze the exertions of the first. Moreover, it will work faultlessly one day, and stall the next. No reliance can be placed upon it at any stated time. Thus it was with the Army when ordered into a general engagement; one corps struggled nobly, whilst the neighboring corps frustrated its efforts by simple inactivity; and whilst the entire Army might fight desperately one day, it would fail in action the following day. Stewart's gallant attack on the 20th was neutralized by Hardee's inertness on the right; and the failure in the battle of the 22d is to be attributed also to the effect of the "timid defensive" policy upon this officer, who, although a brave and gallant soldier, neglected to obey orders, and swing away, totally independent of the main body of the Army.

General Sherman acknowledges the correctness of my position in regard to constant retreat and use of breastworks. He remarks, in reference to the battle of Shiloh:*

> "We did not fortify our camps against an attack, because we had no orders to do so, and because such a course would have made our raw men timid."

* Sherman's Memoirs, vol. II, page 229.

When at Kennesaw Mountain, he ordered General Howard to use freely his artillery, saying: *

"I explained to him that we must keep up the *morale* of a bold offensive, that he must use his artillery, force the enemy to remain on the *timid defensive*."

Again, whilst still at Kennesaw, he says: †

"On the 19th June the rebel Army again fell back on its flanks, to such extent that for a time I supposed it had retreated to the Chattahoochee river. * * * These successive contractions of the enemy's line *encouraged* us and *discouraged* him."

Sherman possessed sufficient judgment and soldiership to discern that the causes which improved his Army, impaired that of his antagonist; and his ground regarding the bold offensive policy in opposition to the "timid defensive," together with his acknowledgment of the effect of breastworks upon raw troops, clearly proves that he did not favor the handling of troops according to the Joe Johnston school.

Lieutenant General S. D. Lee, who served a long period under General Lee, in Virginia, and who was assigned to the command of a corps around Atlanta shortly after I assumed the direction of the Army, remarks in his official report of the offensive operations commencing at Palmetto, Georgia, September 29th, 1864, with reference to the *morale* of the troops during the operations around Atlanta:

"It was my observation and belief that the majority of the officers and men were so impressed with the idea of their inability to carry even temporary breastworks that, when orders were given to attack and there was a probability of encountering works, they regarded it as reckless in the extreme. Being impressed with these convictions, they did not generally move to the attack with that spirit which nearly always ensures success. Whenever the enemy changed his position, temporary works could be improvised in less than two hours, and he could never be caught without them.

---

* Sherman's Memoirs, vol. II, page 53.

† Sherman's Memoirs, vol. II, page 56, italicised by the author.

"In making these observations, it is due to many gallant officers and commands to state that there were noticeable exceptions; but the feeling was so general that anything like a general attack was paralyzed by it. The Army having constantly yielded to the flank movements of the enemy, which he could make with but little difficulty, by reason of his vastly superior numbers, and having failed in the offensive movements prior to the fall of Atlanta, its efficiency for further retarding the progress of the enemy was much impaired; and, besides, the advantages in the topography of the country, south of Atlanta, were much more favorable to the enemy for the movements of his superior numbers than the rough and mountainous country already yielded to him."

Lieutenant General Lee's large experience in Virginia qualified him to form a correct opinion upon this subject; it should also be borne in mind that he assumed command of his Corps around Atlanta, on the 25th of July, immediately after the battles of the 20th and 22d, which had already, in a degree, improved the *morale* of the Army, and which had the subsequent effect of arresting desertions almost entirely throughout the siege.

Notwithstanding my endeavors to explain satisfactorily to myself my inability to procure co-operative action upon the 20th, and 22d, I remained somewhat perplexed upon the subject—especially in regard to the failure, on the 20th, of the best troops of the Army, Hardee's Corps. Shortly after the beginning of the siege, Major General Cleburne, commanding a division in that corps, called at my headquarters. The occurrences of the hour were discussed, and, finally, the two late battles in which he had been a participant. Much was said pro and con, relative to the condition of the Army and the causes of failure in the above referred to engagements. I then unfolded to him the plans of action, together with the peremptory orders to halt at nothing on our side of Peach Tree creek.

Cleburne seemed surprised, and thereupon informed me that as his Division was about to move forward to the attack, on the 20th, General Hardee rode along the line, and, in the presence of those around him, cautioned him to be on the lookout for breastworks.

I can recall no reply on my part at the time, save, perhaps, some expression of astonishment. I could say nothing, even to so worthy a subordinate. He left me to infer, however, from subsequent remarks, that his Division would have taken quite a different action on the 20th, had it not been for the forewarning of his corps commander.

I give the above narrative of facts with a full knowledge of my accountableness to the same Ruler before whom those two gallant soldiers have been summoned; and, as I avowed at the beginning of my task, would not have undertaken to write of these unpleasant subjects, were it not for the seeming perpetuation of injustice and misrepresentation in the guise of truth and history.

It is but reasonable to deduce from this unfortunate observation to Cleburne that General Hardee gave a similar warning to other officers. At all events, those who are able to realize the baneful effect of such a remark from the commander of a corps d'armee, upon the eve of conflict, know that his words were almost equivalent to an order to take no active part in the battle.

From the hour one of the main sources of our trouble was thus accidentally made known to me, I recognized that my power, upon any occasion, to deal quick and heavy blows to the enemy, would be greatly hampered, unless I could procure the relief of this officer and the appointment of one better qualified for the actual emergencies. Whilst General Hardee had, perhaps, no superior as a corps commander during retreat in presence of an enemy, or in defensive operations, he was wanting in that boldness requisite for offensive warfare. This his defect, which may be found in officers of undoubted courage and of every rank, was aggravated by the protracted "timid defensive" policy under my predecessor, and to this misfortune I attributed his non-observance of orders. Long and gallant service had, however, endeared him to his troops, and, because of further demoralization which I feared might ensue in the event of his removal, I decided to retain him in command. Moreover, President Davis held in high appreciation his ability as a corps commander. Lee, Stewart, and G. W. Smith were very open in the expression of their opinion, in regard to his conduct which they imported to a less charitable notice than I was willing to concede. Their opinion of the consequences of his non-fulfilment of orders is recorded in the following extract from the official report of Major General G. W. Smith:

"If they (the corps commanders) are not unanimous, there is but *one*, if any, who dissents from the opinion expressed above, viz: Sherman would have been beaten had your orders been obeyed on the 20th of July, 22d of July, and 31st of August."*

About the Autumn of 1874, I met in St. Louis General Frank Blair, with whom I conversed at length upon military events of the past; and, reverting to the battle of the 22d, I informed him that my instructions to Hardee had been to completely turn McPherson's left, even if he was forced to march to Decatur. He at once remarked that if the move had been accomplished, it would have resulted in the rout of that portion of Sherman's Army; even under the circumstances, the attack nigh proved fatal to the Federal arms.

The following extract from a letter of General Blair to Major J. E. Austin, of New Orleans, who served with great distinction in the Tennessee Army from the beginning to the close of the war, will be read with interest, as the writer commanded a corps in McPherson's Army, during the battle of the 22d of July. This letter was in response to one from Major Austin in relation to different events connected with the Georgia campaign, and touching the two battles under discussion:

"CLIFTON SPRINGS, *February*, 1875.

* * * "Of the affair at Peach Tree creek I know very little, and that only from the report of the officers engaged in it. Our troops there were under the command of General Thomas, who had about fifty thou-

---

* See Report in Appendix, page 354.

sand (50,000) men. Our losses were very severe, and the fighting was very heavy.

"On the 22d of July, my Corps held the extreme left of our Army. We were well entrenched along the McDonough road, running about north and south.

"The reports which we got from the front, early in the morning, indicated some movement of the enemy, and General McPherson, under whom I served, with several other officers and myself rode out to the front to observe what was going on, and, if possible, to make up our minds as to the nature of the movement which was being made. We approached the city near enough to see plainly that a large body of troops were moving out of the city towards the south, and great numbers of the citizens, including some of the ladies, were on the tops of the houses as if they were expecting some movement to take place which they were desirous of seeing; yet the Confederate entrenchments, immediately in front of us were full of men, who, however, did not fire upon us although we were very near them, and in plain view. McPherson said that he believed that the enemy were abandoning the city and were in full retreat, and that the citizens were on the tops of the houses expecting to see our Army enter the city. As we rode back to our lines, General McPherson repeatedly expressed his opinion that the Confederates were retreating, and would abandon the city to us. I rode with him to his headquarters, and, after I had been there some half hour, we heard skirmishing in our rear, immediately in the direction of Decatur. General McPherson ordered me to send back to the rear two regiments to protect our hospitals. I executed this order, and, as these regiments were moving to the rear, one division of the Sixteenth Corps, which had been ordered by General McPherson to take position on my extreme left, made its appearance in rear of my position, on a road known as the Clay road, and at right angles with the McDonough road, along which my Corps was entrenched.

"Just as this division halted, about five hundred yards in rear of my line, heavy skirmishing commenced on the extreme left of my line. I hastened toward the front of the line, and as I reached a skirt of timber which intervened between me and the line, I saw General McPherson, accompanied by one orderly, enter this piece of timber in front of me. In a few minutes I heard a heavy discharge of musketry, and McPherson's horse came out of the timber riderless.

"The division of the Sixteenth Corps, to which I have alluded, in a very few moments became heavily engaged, and I became aware of the fact that my whole position had been turned, and that the enemy were pressing with full force upon the rear and flank of my position. General McPherson had been killed in attempting to reach my line, on a road over which we had ridden away from that line a short time before, in the full belief that the enemy were in retreat.

"I was only able to reach the line by making a detour to the right, and reached it at a point where it joined the Fifteenth Corps, to find the whole of my line fighting from the reverse of my entrenchments.

"The Confederates were very much scattered, and, I dare say, fatigued by their long and swift march, and did not make a very vigorous attack. The diversion created by the division of the Sixteenth Corps was also a very great assistance to us. We had hardly got rid of the attack in our rear before we were assailed from the direction of Atlanta, but this attack was easily repulsed in my front, although it was more successful on the front of the Fifteenth Corps which was broken, and driven from its entrenchments by a large body of Confederates who had collected in the rear of a large fine house, which had been allowed to stand, a short distance in front of our line. It stood on the main road from Decatur to Atlanta, and for some reason, had not been destroyed as it should have been. A large body of men had collected in the rear of this house, and, when this attack was made, they precipitated themselves on the line of the Fifteenth Corps, driving them from their entrenchments.

"When I saw that the Fifteenth Corps had been driven from their entrenchments, knowing that the position of my Corps had been completely turned, I was convinced that I should not be able to maintain my position; but the Fifteenth Corps rallied gallantly, and recovered their lost ground.

"Although the attack upon us was renewed again and again, both from the front and rear, we were still able to maintain our position. Late in the day, I drew out my forces from the line which they had occupied, and took up a new position, extending from the hill where my right had formerly rested, and extending toward the position in which I have described the Sixteenth Corps to have occupied in my rear. This new position prevented the enemy from taking me in the rear.

"We had barely time to throw up a very tight rifle pit, before the enemy attacked us with great vigor in our new position, and, when night closed in upon us, the fighting still continued; and the lines were so close that it was impossible for a person looking on to tell one line from the other, except for the direction of the fire from the muzzles of their guns.

"On the next morning at 10 o'clock, we had a truce for burying the dead. As we had given up the greater part of the ground over which the battle had been fought the day before, most of our dead were within their lines. We had suffered very severely: we had lost many valuable officers, including General McPherson, but, as we had fought from behind entrenchments all the time, the Confederate loss had necessarily been much greater than ours; and as the dead were separated into different piles by the working parties who were sent out from each Army, the difference was very striking and must have been observed by every one.

"The movement of General Hood was a very bold and a very brilliant one, and was very near being successful.

"The position taken up accidentally by the Sixteenth Corps prevented the full force of the blow from falling where it was intended to fall. If my command had been driven from its position at the time that the Fifteenth Corps was forced back from its entrenchments, there must have been a general rout of all the troops of the Army of the Tennessee, commanded by General McPherson, and, possibly, the panic might have been communicated to the balance of the Army. This, however, is not likely, as Thomas's command and Schofield's together, made a much larger force than the whole Army of Hood, and they were not easily put into panic. As it was, we congratulated ourselves on being able to hold our position, and we felt satisfied that Hood's Army could not stand much longer the terrible losses it was suffering from these brilliant but disastrous movements. The opinion in our Army was that the result would have been the same if Joe Johnston had continued in command, but that the denouement was hastened and expedited by the change of tactics adopted by General Hood. This I think, and indeed am sure, was General Sherman's opinion before and after Hood's tactics were put in practice.

"I remember to have got a newspaper from a farm house, in which the change of commanders was announced. I got it on the very morning it was printed, and sent it immediately to General Sherman by one of my couriers. He wrote me back that it was very good news, but to look out for an attack; that Hood would make it very lively for us, and that it was necessary to be exceedingly cautious.

"I don't know of anything that I have in my power to say now, which will throw any light on the subject of your inquiry, but, in reply to your second question, I would say that I do not believe our Army ever undertook, or attempted, any flanking or turning operations without using entrenchments; at least I have no recollection of their ever doing so.

"In conclusion, I cannot help expressing regret that any misunderstanding should have occurred between two such gallant officers as General Hood and General Johnston, and their friends. Both of them were most meritorious officers and commanded the respect and admiration of their enemies. The great fault of both was that they did not have men enough to contend with Sherman's Army. It was natural enough that after the failure of General Johnston to check our advance, other tactics should be employed; and no man could have been found who could have executed this policy with greater skill, ability and vigor than General Hood.

"With many thanks for your kind expressions towards me personally,

"I remain, your friend,

"FRANK P. BLAIR.

"To Colonel J. E. Austin, *New Orleans.*"

General Blair was mistaken in pronouncing the attack *disastrous*, since, as I have stated, it greatly improved the *morale* of the Army, and arrested desertion. In connection with the battle of the 20th, it also enabled us to hold possession of Atlanta a prolonged period. He erred likewise in attributing the lack of spirit in Hardee's troops to fatigue from the march of the night previous. Decatur is but six miles from Atlanta, and the detour required to be made was but slight. Beside, those troops had been allowed almost absolute rest the entire day of the 21st.

Stonewall Jackson made a hard march, in order to turn Pope at Second Manassas, and again to come up in time at Antietam, or Sharpsburg; as also at Chancellorsville, in order to fall upon Hooker's flank and rear. Longstreet likewise made hard marches, prior to the battles of Second Manassas and Gettysburg. The men were often required, under Lee, to perform this kind of service an entire day and night, with only a halt of two hours for sleep, in addition to the ordinary rests allowed on a march; and were then expected to fight two or three consecutive days. Indeed, in movements of this character, it is rare that a decided advantage is gained over an enemy, without the endurance of great fatigue and privation on the part of the troops. Neither Johnston's nor Sherman's Armies ever experienced the weariness and hardship to which Lee and Jackson frequently subjected their troops—the fruits of which, brought to perfection by their transcendent genius, won for them a fulness of glory and renown, shared by no other soldier of the war.

I am as thoroughly convinced at present as at the hour these events transpired, that had these same forces, at my disposal in these battles, been previously handled according to the Lee and Jackson school, they would have routed the Federal Army, and, in all probability have so profited by Sherman's blunders as to have altered signally the issue of these operations.

CHAPTER XII.

SIEGE OF ATLANTA—ENGAGEMENT OF THE 28TH OF JULY—WHEELER, IVERSON AND JACKSON—BATTLE OF JONESBORO'—EVACUATION OF ATLANTA.

In accordance with the valuable diary of Brigadier General Shoupe, I find naught to record after the battle of the 22d beyond the usual shelling by the enemy, till the 26th of July when the Federals were reported to be moving to our left. This movement continued during the 27th, when I received the additional information that their cavalry was turning our right, in the direction of Flat-rock, with the intention, as I supposed, of interrupting our main line of communication, the Macon Railroad. We had lost the road to Augusta previous to the departure of General Johnston on the 18th, and, by the 22d, thirty miles or more thereof had been utterly destroyed.

The Federal commander continued to move by his right flank to our left, his evident intention being to destroy the only line by which we were still able to receive supplies. The railroad to West Point, because of its proximity to the Chattahoochee river, was within easy reach of the enemy whenever he moved far enough to the right to place his left flank upon the river. Therefore, after the destruction of the Augusta road, the holding of Atlanta—unless some favorable opportunity offered itself to defeat the Federals in battle—depended upon our ability to hold intact the road to Macon.

Sherman thus refers to the importance of this line:*

> "I always expected to have a desperate fight to get possession of the Macon road, which was then the vital objective of the campaign."

General Wheeler started on the 27th of July in pursuit of the Federal cavalry which had moved around our right; and General Jackson, with the brigades of Harrison and Ross, was ordered, the following day, to push vigorously another body of the enemy's cavalry which was reported to have crossed the river, at Campbellton, and to be moving, via Fairburn, in the direction of the Macon road. On the 28th it was apparent that Sherman was also moving in the same direction with his main body. Lieutenant General Lee was instructed to move out with his Corps upon the Lick-Skillet road, and to take the position most advantageous to prevent or delay the extension of the enemy's right flank. This officer promptly obeyed orders, and came, unexpectedly, in the afternoon, in contact with the Federals in the vicinity of Ezra Church, where a spirited engagement ensued. The enemy was already in possession of a portion of the ground Lee desired to occupy, and the struggle grew to such dimensions that I sent Lieutenant General Stewart to his support. The contest lasted till near sunset without any material advantage having been gained by either opponent. Our troops failed to dislodge the enemy from their position, and the Federals likewise to capture the position occupied by the Confederates. Although the actual loss was small in proportion to the numbers engaged, Generals Stewart, Brown, Loring, and Johnson, were slightly wounded. I desired of Lieutenant General Lee an opinion as to the manner in which our troops had conducted themselves upon the field. In answer to my request, he replied that he could not succeed in bringing about united action; whilst one

---

*Sherman's Memoirs, vol. II, page 99.

brigade fought gallantly, another failed to do its duty. I learned afterwards that such indeed was the case, notwithstanding he had led one or more to the attack, and had even offered to lead others. Although this affair occurred subsequent to the improvement of the *morale* of the Army and the check to desertions, which had resulted from the battles of the 21st and 22d, the lack of spirit manifested in this instance will convey a just idea of the state of the Army at this period.

In reference to the non-capture of the position held by the enemy, he says in his official report:

> "I am convinced that if all the troops had displayed equal spirit, we would have been successful, as the enemy's works were slight, and, besides, they had scarcely gotten into position when we made the attack."

Whilst these operations were in progress, Wheeler and Jackson were in hot pursuit of the Federal cavalry—General Lewis's infantry brigade having been sent to Jonesboro', the point about which I supposed the raiders would strike our communications.

At an early hour on the 29th, dispatches were received from various points upon the Macon road to the effect that General Wheeler had successfully checked the enemy at Latimer's, and was quietly awaiting developments. On our left, the Federals succeeded in eluding our cavalry, for a time, by skirmishing with our main body, whilst their main force moved round to the rear, and cut the telegraph lines at Fairburn and Palmetto. General Jackson, however, soon discovered the ruse, and marched rapidly toward Fayetteville and Jonesboro', the direction in which the Federals had moved. The enemy succeeded in destroying a wagon train at the former place; in capturing one or two quarter masters who afterwards made their escape, and in striking the Macon road about four miles below Jonesboro', when the work of destruction was began in earnest.

General Lewis, within three hours after receiving the order, had placed his men on the cars and was in Jonesboro' with his brigade, ready for action. Meantime Jackson was coming up with his cavalry, when the Federals became alarmed, and abandoned their work; but not without having destroyed about a mile and a-half of the road, which was promptly repaired.

While Jackson followed in pursuit, and Lewis returned to Atlanta, Wheeler moved across from Latimer's, with a portion of his command, in rear of this body of the enemy—leaving General Iverson to pursue General Stoneman who, after somewhat further damaging the Augusta road, and burning the bridges across Walnut creek and the Oconee river, had moved against Macon.

These operations had been ordered by General Sherman upon a grand scale; picked men and horses had been placed under the command of Generals McCook and Stoneman, with the purpose to destroy our sole line of communication, and to release, at Andersonville, thirty-four thousand (34,000) Federal prisoners to ravage and pillage the country.

These raiders, under McCook, came in contact with General Roddy's cavalry at Newnan, and were there held in check till Wheeler's and Jackson's troops came up; whereupon the combined forces, directed by General Wheeler, attacked the enemy with vigor and determination, and finally routed them. Whilst these operations were progressing in the vicinity of Newnan, General Cobb was gallantly repelling the assault of Stoneman at Macon, when Iverson came up, and engaged the enemy with equal spirit and success.

The following dispatches were received from Generals Wheeler and Iverson. Wheeler says:

"We have just completed the killing, capturing, and breaking up of the entire raiding party under General McCook—some nine hundred and fifty (950) prisoners—two pieces of artillery, and twelve hundred horses and equipments captured."

Iverson, the same date:

"General Stoneman, after having his force routed yesterday, surrendered with five hundred (500) men; the rest of his command are scattered and flying toward Eatonton. Many have been already killed and captured."

General Shoupe, in recording these two telegrams in his diary, states that Iverson also captured two pieces of artillery, and remarks that "the 1st day of August deserves to be marked with a white stone." He, doubtless in common with every Southerner, experienced deep concern in regard to the Federal prisoners at Andersonville, as it was reported that Sherman had arms in readiness for their use. Fearful indeed would have been the consequences, had they been turned loose upon the country in its unprotected condition.

Had the authorities at Richmond believed that General Johnston would have abandoned the strongholds of the mountains, they would assuredly have removed these prisoners before the Federals crossed the Chattahoochee.

General Sherman, in reference to his plan of operations at this time, writes: *

"My plan of action was to move the Army of the Tennessee to the right rapidly and boldly against the railroad below Atlanta, and at the same time to send all the cavalry round by the right and left to make a lodgment on the Macon road about Jonesboro'."

The flanks of the Federal Army were at this juncture so well protected by the Chattahoochee and the deep ravines which run down into the river, that my antagonist was enabled to throw his entire force of cavalry against the Macon road; and but for the superiority of the Confederate cavalry, he might have succeeded to such extent as to cause us great annoyance, and subject our troops to short rations for a time.

After the utter failure of this experiment, General Sherman perceived that his mounted force, about twelve thousand in number, in concert with a corps of infantry as support, could not so effectually destroy our main line of communication as

* Sherman's Memoirs, vol. II, page 87.

to compel us to evacuate Atlanta, as the subjoined extract will indicate:*

"I now became satisfied that cavalry could not, or would not, make a sufficient lodgment on the railroad below Atlanta, and that nothing would suffice but for us to reach it with the main Army."

Wheeler and Iverson having thus thoroughly crippled the Federal cavalry, I determined to detach all the troops of that arm I could possibly spare, and expedite them, under the command of Wheeler, against Sherman's railroad to Nashville; at the same time, to request of the proper authorities that General Maury, commanding at Mobile, be instructed to strike with small bodies the line at different points, in the vicinity of the Tennessee river, and also that General Forrest be ordered with the whole of his available force into Tennessee for the same object. I intended General Wheeler should operate, in the first instance, south of Chattanooga.

I was hopeful that this combined movement would compel Sherman to retreat for want of supplies, and thus allow me an opportunity to fall upon his rear with our main body. I expressed this hope in a dispatch of August 2d, to President Davis. In reply thereto, and I presume also to a letter indited the ensuing day, but of which I possess no copy, he sent the following telegram:

"RICHMOND, *August 5th, 1864*.

"GENERAL J. B. HOOD.

"Yours of August 3d received. I concur in your plan, and hope your cavalry will be able to destroy the railroad bridges and depots of the enemy on the line to Bridgeport, so as to compel the enemy to attack you in position or to retreat. The loss consequent upon attacking him in his entrenchments requires you to avoid that if practicable. The enemy have now reached a country where supplies can be gathered by foraging expeditions, and a part of your cavalry will be required to prevent that. If he can be forced to retreat for want of supplies, he will be in the worst condition to escape or resist your pursuing Army. General Hardee's minute knowledge of the country and his extensive acquaint-

---

*Sherman's Memoirs, vol. II, page 98.

ance with the officers and men of the command, must render his large professional knowledge and experience peculiarly valuable in such a campaign as I hope is before you.

"JEFFERSON DAVIS."

The foregoing dispatch is the only communication offering a suggestion, which I remember to have received during the siege of Atlanta from the President; it therefore stands out in bold contradiction to the general assertion that I was ordered by him to assume the offensive, or to make certain campaigns. The President did not, at any time, order what I should or should not do; and although I had solicited counsel, he gave none, save the above caution in regard to breastworks, and, at a later period, his expressed disapproval of the contemplated campaign into Tennessee.

In accordance with my determination to attempt, with cavalry, the destruction of Sherman's road, I ordered General Wheeler with four thousand five hundred (4500) men to begin operations at once. He succeeded in burning the bridge over the Etowah; recaptured Dalton and Resaca; destroyed about thirty-five miles of railroad in the vicinity, and captured about three hundred mules and one thousand horses; he destroyed, in addition, about fifty miles of railroad in Tennessee.

General Forrest, with his usual energy, struck shortly afterwards the Federal line of supplies in this State, and, as will hereafter be shown, inflicted great damage upon the enemy. Of his exploits on this expedition I have no official report, as he was not directly under my command.

Forrest and Wheeler accomplished all but the impossible with their restricted number of cavalry, and the former, finally, was driven out of Tennessee by superior forces. General Sherman, in relation to this movement, says: *

"The rebel General Wheeler was still in Middle Tennessee, threatening our railroads, and rumors came that Forrest was on his way from

* Sherman's Memoirs, vol. II, page 130.

Mississippi to the same theatre, for the avowed purpose of breaking up our railroads and compelling us to fall back from our conquest. To prepare for this, or any other emergency, I ordered Newton's Division of the Fourth Corps back to Chattanooga, and Corse's Division of the Seventeenth Corps to Rome, and instructed General Rosseau at Nashville, Granger at Decatur, and Stedman at Chattanooga, to adopt the most active measures to protect and insure the safety of our roads."

So vast were the facilities of the Federal commander to reinforce his line of skirmishers, extending from Nashville to Atlanta, that we could not bring together a sufficient force of cavalry to accomplish the desired object. I thereupon became convinced, and expressed the opinion in my official report, that no sufficiently effective number of cavalry could be assembled in the Confederacy to interrupt the enemy's line of supplies to an extent to compel him to retreat.

From the 5th to the 19th of August no event of special importance occurred. I find naught recorded save the constant demonstrations of the enemy in front, whilst completing his movement to our left. A heavy demonstration was made on the 6th against Bates's Division which was twice assaulted; twice the foe were driven back in great confusion with a loss of two stands of colors, eight hundred killed and wounded, some small arms and entrenching tools.

On the 7th General Cleburne's Division was transferred to our extreme left, and the 9th was made memorable by the most furious cannonade which the city sustained during the siege. Women and children fled into cellars, and were there forced to seek shelter a greater length of time than at any period of the bombardment.

The 19th, nigh two weeks after Wheeler's departure with about one-half of our cavalry force, General Sherman took advantage of the absence of these troops, and again attempted a lodgment on the Macon road with cavalry. At 3.30 a. m., General Kilpatrick was reported to be moving, via Fairburn, in the direction of Jonesboro'. General Jackson quickly divined his object, moved rapidly in pursuit, overtook him at an early hour, attacked and forced him to retreat after sustaining considerable loss in killed, wounded and prisoners. The Federals had previously succeeded, however, in destroying a mile and a-half of the Macon road; had cut the wires, and burned the depot at Jonesboro'.

General Sherman, touching this his second strenuous effort to render the evacuation of Atlanta a matter of compulsion by throwing cavalry to our rear, says: *

"He (Kilpatrick) reported that he had destroyed three miles of the railroad about Jonesboro', which he reckoned would take ten days to repair; that he had encountered a division of infantry and a brigade of cavalry (Ross's); that he had captured a battery and destroyed three of its guns, bringing one in as a trophy, and he also brought in three battle-flags and seventy prisoners. On the 23d, however, we saw trains coming into Atlanta from the South, when I became more than ever convinced that cavalry could not or would not work hard enough to disable a railroad properly, and therefore resolved at once to proceed to the execution of my original plan."

Our cavalry also drove a brigade of the enemy from the Augusta road on the 22d, which affair, together with the happy results obtained in the engagement with Kilpatrick, demonstrated conclusively that the absence of one-half of our mounted force notwithstanding, we had still a sufficient number, with Jackson, to protect not only the flanks of the Army, but likewise our communications against similar raids, and, moreover, to defend our people against pillaging expeditions. At this period, I was charged by the Johnston-Wigfall party, through the press, with having committed a serious blunder by sending off the cavalry, and with having exposed our people to robbery and maltreatment by raiders through the country.

The severe handling by Wheeler and Iverson of the troops under Stoneman and McCook, together with Jackson's success, induced me not to recall Wheeler's four thousand five hundred (4500) men who were still operating against the railroad to

* Sherman's Memoirs, vol. II, page 104.

Nashville. I had, moreover, become convinced that our cavalry was able to successfully compete with double their number. Fortunately, they had not become demoralized upon the retreat, in consequence of their habit of dismounting and fighting at one point to-day, then remounting and hastening in another direction to encounter the enemy on the morrow. As before stated, our cavalry were not cavalrymen *proper*, but were mounted riflemen, trained to dismount and hold in check or delay the advance of the main body of the enemy, and who had learned by experience that they could without much difficulty defeat the Federal cavalry. This teaching, combined with the fact that small bodies can fall back in front of large armies without material discouragement to the men, warded off the baneful influences which worked upon the infantry, and accounts for the non-demoralization of the cavalry.

In this connection, it becomes my duty, as well as pleasure, to make acknowledgments of the valuable services of the cavalry of the Army of Tennessee, during my operations in Georgia, and North Alabama. I have not forgotten the outcry against Wheeler's cavalry just prior to and after the close of the war; it was brought about in great measure, doubtless, by renegades from our Armies, who committed outrages which were charged by the people to the account of the cavalry. I am confident that when the history of our struggle is written, Major General Wheeler and his command will occupy a high position, as the Confederacy possessed, in my opinion, no body of cavalry superior to that which I found guarding the flanks of the Army of Tennessee at the time I assumed its direction.

The bombardment of the city continued till the 25th of August; it was painful, yet strange, to mark how expert grew the old men, women and children, in building their little underground forts, in which to fly for safety during the storm of shell and shot. Often 'mid the darkness of night were they constrained to seek refuge in these dungeons beneath the earth; albeit, I cannot recall one word from their lips, expressive of dissatisfaction or willingness to surrender.

Sherman had now been over one month continuously moving toward our left and thoroughly fortifying, step by step, as he advanced in the direction of the Macon Railroad. On the night of the 25th, he withdrew from our immediate front; his works, which at an early hour the following morning we discovered to be abandoned, were occupied at a later hour by the corps of Generals Stewart and Lee.

This movement of the Federals gave rise to many idle rumors in relation to its object. I felt confident that their plan would soon be developed; accordingly, orders were issued to corps commanders to send out scouts in their front, and to keep Army headquarters fully advised of the slightest change in the enemy's position; to issue three days' rations, and to be in readiness to move at a moment's warning. Instructions were likewise sent to General Armstrong, commanding the cavalry in the vicinity of the West Point Railroad, to be most active in securing all possible information in regard to the operations of the enemy.

On the 27th, Major General G. W. Smith's Division was ordered to the left to occupy the position of Stevenson's Division which, together with General Maury's command, was held in reserve. Early the following morning, the enemy were reported by General Armstrong in large force at Fairburn, on the West Point road. It became at once evident that General Sherman was moving with his main body to destroy the Macon road, and that the fate of Atlanta depended upon our ability to defeat this movement.

Reynolds's and Lewis's brigades were dispatched to Jonesboro' to co-operate with Armstrong. General Adams, at Opelika, was directed to guard the defences of that place with renewed vigilance, while General Maury was requested to render him assistance, if necessary. The chief quarter master, ordnance officer, and commissary, were given most explicit instructions in regard to the disposition of their respective stores. All surplus property, supplies, etc., were ordered to the rear, or to be placed on cars in readiness to move at any moment the railroad became seriously threatened. General Armstrong was instructed to establish a line of couriers to my headquarters, in order to report every hour, if requisite, the movements of the enemy. In fact, every precaution was taken not only to hold our sole line of communication unto the last extremity, but also, in case of failure, to avoid loss or destruction of stores and material.

On the 29th, the Federals marched slowly in the direction of Rough and Ready, and Jonesboro'. A portion of Brown's Division was directed to take position at the former place and fortify thoroughly, in order to afford protection to the road at that point. General Hardee, who was at this juncture in the vicinity of East Point, was instructed to make such disposition of his troops as he considered most favorable for defence; and, in addition, to hold his Corps in readiness to march at the word of command. Generals Jackson and Armstrong received orders to report the different positions of the corps of the enemy at dark every night.

Had Sherman not been doubly protected by the Chattahoochee, deep intervening creeks and ravines extending to the river, beside the wall of parapets behind which he had thus far manœuvred, I would have moved out from East Point with our main body, and have attacked his Army whilst effecting these changes of position. This move not being practicable by reason of these obstructions, I was forced to await further developments.

The morning of the 30th found our general line extended further to the left—Hardee being in the vicinity of Rough and Ready with Lee's Corps on his right, near East Point. Information from our cavalry clearly indicated that the enemy would strike our road at Jonesboro'; after consultation with the corps commanders, I determined upon the following operations, as the last hope of holding on to Atlanta.

As General Armstrong had already foreseen, a Federal corps

crossed Flint river at about 6 p. m., near Jonesboro', and made an attack upon Lewis's brigade, which was gallantly repulsed. This action became the signal for battle. General Hardee was instructed to move rapidly with his troops to Jonesboro', whither Lieutenant General Lee, with his Corps, was ordered to follow during the night. Hardee was to attack with the entire force early on the morning of the 31st, and drive the enemy, at all hazards, into the river in their rear. In the event of success, Lee and his command were to be withdrawn that night back to Rough and Ready; Stewart's Corps, together with Major General G. W. Smith's State troops were to form line of battle on Lee's right, near East Point, and the whole force move forward the following morning, attack the enemy in flank, and drive him down Flint river and the West Point Railroad. In the meantime, the cavalry was to hold in check the corps of the enemy, stationed at the railroad bridge across the Chattahoochee, near the mouth of Peach Tree creek, whilst Hardee advanced from his position near Jonesboro', or directly on Lee's left.

Such were the explicit instructions delivered. I impressed upon General Hardee that the fate of Atlanta rested upon his ability, with the aid of two corps, to drive the Federals across Flint river, at Jonesboro'. I also instructed him, in the event of failure—which would necessitate the evacuation of the city—to send Lee's Corps, at dark, back to or near Rough and Ready, in order to protect our retreat to Lovejoy Station.

I remained in Atlanta with Stewart and G. W. Smith, anxiously awaiting tidings from Jonesboro'. At an early hour the following morning, no information having been received and the wires having been cut by the enemy, I despatched a courier with orders that Lee's Corps, in any event, march back and take position in the vicinity of Rough and Ready. The arrival of no messenger from Hardee caused me to fear that the attack had not been made at an early hour, according to instructions; this apprehension proved, unfortunately, but too well grounded.

The attack was not made till about 2 p. m., and then resulted in our inability to dislodge the enemy. The Federals had been allowed time, by the delay, to strongly entrench; whereas had the assault been made at an early hour in the morning, the enemy would have been found but partially protected by works. Lieutenant General S. D. Lee expressed the opinion, at the time, that the enemy could have been driven across the river, if the attack had been made at an early hour, or soon after his Corps arrived at Jonesboro'. General Hardee transmitted to me no official report at that period, nor subsequently, of his operations whilst under my command. I find, however, from the diary in my possession that his Corps succeeded in gaining a portion of the Federal works; the general attack, notwithstanding, must have been rather feeble, as the loss incurred was only about fourteen hundred (1400) in killed and wounded—a small number in comparison to the forces engaged. Among the wounded were Major General Patton Anderson and Brigadier General Cummings, who were disabled whilst gallantly leading their troops into action.

This failure gave to the Federal Army the control of the Macon road, and thus necessitated the evacuation of Atlanta at the earliest hour possible.

I was not so much pained by the fall of Atlanta as by the *recurrence* of retreat, which I full well knew would further demoralize the Army and renew desertions. The loss of over four thousand (4000), sustained from this same cause during the change from Kennesaw Mountain to and across the Chattahoochee, augmented my great reluctance to order the Army to again turn its back to the foe. Howbeit, as stated in my official report, the presence of thirty-four thousand (34,000) Federal prisoners at Andersonville, rendered it absolutely incumbent to place the Army between Sherman and that point, in order to prevent the Federal commander from turning loose this large body, ready to wreak its ill-will upon our people. Thus the proximity of these prisoners to Sherman's Army not only forced me to remain in a position to guard the country against the fearful calamity aforementioned, but also thwarted my design to move north, across Peach Tree creek and the Chattahoochee, back to Marietta, where I would have destroyed the enemy's communications and supplies, and then have taken position near the Alabama line, with the Blue Mountain Railroad in rear, by which means the Confederate Army could, with ease, have been provisioned.* Notwithstanding the presence of one of Sherman's Corps at the railway bridge over the Chattahoochee, I would have made this move. I would have thrown upon our left flank a sufficient force to occupy the Federals, at the bridge, whilst we laid pontoons and passed round to their rear, as we subsequently did in the presence of Schofield, at Columbia, Tennessee. Had I been enabled to carry into effect this plan, Hardee and Lee would not have been sent to Jonesboro', as the cavalry would have been instructed to retard, to the utmost, the advance of the enemy, whilst Major General Cobb made demonstrations from the direction of Macon. Thus, while Sherman was destroying the road to Macon, I would have been upon his communications with Nashville, and the desertions, together with the demoralization which followed the evacuation of Atlanta, would have been avoided.

In lieu of the foregoing operations, the battle of Jonesboro' was fought, and on the following day, September 1st, at 2 a.m., Lieutenant General Lee, with his Corps, marched from Jonesboro' to the vicinity of Rough and Ready; and so posted his troops as to protect our flank, whilst we marched out of Atlanta at 5 p. m. the same day, on the McDonough road, in the direction of Lovejoy Station. Generals Morgan and Scott, stationed at East Point, received similar orders to protect our flank during the retreat.

Upon our uninterrupted march, information reached me that Hardee's Corps was engaged with a large force of the enemy. His position upon a ridge with an open country in rear relieved

---

* See Official Report, Appendix page 324.

me from special anxiety in regard to the safety of himself and command. Lieutenant General Stewart, nevertheless, was instructed to hasten forward to his support, and General Lee to follow promptly with his Corps. When these reinforcements reached the scene of action, the contest had ceased. Hardee's troops had been attacked by a considerable force; but, in consequence of the protection afforded by their breastworks, their loss in killed and wounded was small in comparison to that of the enemy. The Federals, who largely exceeded them in numbers, forced them back a short distance from the position they primarily occupied, and necessitated the abandonment of two four gun batteries. This engagement was the only event of importance which occurred during our continuous march from Atlanta to Lovejoy Station. I have often thought it strange Sherman should have occupied himself with attacking Hardee's entrenched position, instead of falling upon our main body on the march round to his rear.

Notwithstanding full and positive instructions, delivered prior to the evacuation of the city, and ample time and facilities afforded to move all stores, cars and engines, the chief quarter master grossly neglected to send off a train of ordnance stores, and five engines, although they were on the track and in readiness to move. This negligence entailed the unnecessary loss of these stores, engines, and about eighty cars. Shortly afterwards, a Court of Inquiry was assembled to examine into and report upon the cause of this unwarranted loss. A copy of the findings of the Court was forwarded to Richmond, and is at present, I presume, among the captured Confederate records in Washington. I regret I possess no copy for reference; my memory, however, is quite clear as to the result of the inquiry, which was, in substance, that the commanding general had issued all necessary orders, and was in no manner to blame; that ample time had been allowed, and the road left open long enough to have transferred all stores, etc., to a place of safety in rear; and that the loss was to be attributed to the neglect of the chief quarter master. The stores which had been abandoned were blown up at about 2 o'clock on the morning of the 2d September, and the rear guard soon thereafter marched out of Atlanta. That night and the morning of the 3d, our troops filed into position in Sherman's front, which was then near Jonesboro'. By the 4th, our entire Army was assembled at this point, on the Macon road.

Major General Gustavus W. Smith, commanding Georgia State troops, was directed to proceed to Griffin and protect our communications in that vicinity; General Jackson was ordered to keep active scouts in the direction of Greenville; General Morgan to report to Jackson for duty; Lewis's Kentucky brigade to be mounted, and to use blankets in default of saddles.

On the 5th, General Morgan was ordered back to assume command of the cavalry on the right; the corps commanders were instructed to use every effort to gather up absentees; the chief commissary was directed to keep on hand five days' rations of hard bread; Major Beecher, quarter master, to confer with Major Hallett, superintendent of the railroad, in regard to means to facilitate the transportation of supplies, and to issue shoes and clothing forthwith upon their receipt.

On the 6th, the Federals withdrew from our immediate front, and moved off in the direction of Atlanta. General Sherman published orders stating that his Army would retire to East Point, Decatur, and Atlanta, and repose after the fatigue of the campaign through which it had passed. We were apprised of these instructions soon after their issuance—as well as of nigh every important movement of the enemy through the vigilance of our cavalry, spies, and scouts, and from information received through Federal prisoners. Upon this date it may be justly considered that the operations round Atlanta ceased. We had maintained a defence, during forty-six days, of an untenable position, and had battled almost incessantly, day and night, with a force of about forty-five thousand (45,000) against an Army of one hundred and six thousand (106,000) effectives, flushed with victory upon victory from Dalton to Atlanta. When we recall the extent of the demoralization of the troops at the commencement of the siege, we cannot but recognize that the Army of Tennessee was composed of splendid material, and that its condition at Dalton justifies the assertion of its capability, by proper handling, of having been made the equal of its counterpart in Virginia. The non-fulfilment of its brilliant promise is nowise attributable to the officers and men. The fault lies at the door of the teacher in whose school they had been trained; therefore, none of my countrymen can hesitate to accord them the highest praise for the patriotic and noble work performed by them during the siege of Atlanta.

CHAPTER XIII.

ATLANTA UNTENABLE—LOSSES DURING THE SIEGE COMPARED WITH THOSE OF SHERMAN, AND WITH THOSE OF JOHNSTON FROM DALTON TO ATLANTA.

HAVING stated that our position at Atlanta was untenable, I shall now undertake to give proof of the correctness of my opinion, by demonstrating in what manner Sherman might have captured the city in less than one-third of the time he actually devoted to that end, notwithstanding the idle assertion of General Johnston that he could have held Atlanta "forever"; also to demonstrate that had I ventured to remain longer than fifteen days within the trenches around the city, Sherman could have finally forced me to surrender, or have put my Army to rout in its attempt to escape; and, lastly, that he could have attained his object with one-fourth the loss he sustained during a siege of forty-six days, provided he had availed himself of the natural advantages afforded him.

In order to render these operations clear to the mind of the reader, I invite his attention to a map, page 167.

The Federal commander chose well his crossing of the Chattahoochee, as he approached Atlanta, and the move of McPherson and Schofield upon the Augusta road was ably conceived and executed. Thomas, however, should not have formed line of battle along the lower part of Peach Tree creek with a view to cross the creek, as he endangered the safety of his Corps when he placed it in the angle, formed by this stream and the Chattahoochee, and thus isolated himself from Schofield and McPherson. His right should have rested in the vicinity of, and have covered, the ford nearest the mouth of Peach Tree creek, with a line of skirmishers extending to the Chattahoochee, and batteries in position at proper intervals in rear of this line, and likewise on the north side of the river, so as to thoroughly command the approaches to the railway bridge. His right being established at this ford, his left should have been thrown back north of Decatur, and his entire line strongly entrenched. From this position of perfect safety, he could have made constant demonstrations against the city, whilst McPherson and Schofield destroyed the road to Augusta. At the same time, by use of the batteries near the mouth of Peach Tree creek, and of those on the north bank of the river, near the railway bridge, he could have easily thrown a division across the creek, and have established a strong *tête-de-pont* on the south bank of the river. I would thus have been forced to form line of battle facing Peach Tree, with no possible chance of successfully assaulting the enemy at any point. His right and rear would have been covered by a deep and muddy stream; his front protected not only by breastworks, but also by one of the branches of Peach Tree creek. I could not have attacked either his left or McPherson and Schofield, without marching out of Atlanta, and exposing our left flank to Thomas. I would have been compelled to abide quietly the destruction of one of our principal roads, without the ability to strike a blow in its defence.

After my loss of the Augusta road, McPherson and Schofield should have marched by the right flank down Peach Tree, in rear of Thomas's line, until their right rested on the Chattahoochee, and then have halted. General Sherman having his Army thus massed, and well in hand, in rear of Peach Tree creek, should have thrown across the Chattahoochee a sufficient number of pontoon bridges to allow the easy and rapid passage of his troops; have sent two corps of Thomas's Army across and down the Chattahoochee, on the northwest side, to a favorable crossing just below Camp creek or one of the deep ravines or creeks, heading in the direction of East Point, and running toward the river upon the southwest side; have laid pontoons, crossed over, and strongly entrenched. Whilst this move was in progress, the main body should have made heavy demonstrations along the line of Peach Tree to the Augusta road, which diversion would have held my Army in position on the north side of Atlanta.

The two corps below Camp creek having their line and outposts established sufficiently in advance to allow full space for the massing of the Army, Thomas should have left the division in the *tête-de-pont* protected by ditch and abatis, whilst another division, with dismounted cavalry, occupied his position on Peach Tree creek with cavalry on their left, and a few batteries to support their line; then have marched, at dusk, with the remainder of his Army to join the two corps below Camp creek, followed by Schofield and McPherson.

The transportation of the Federal Army having been previously parked on the north side of the Chattahoochee, General Sherman could with entire safety have massed his Army in the space of one night on the southeast side of the river, below this creek, as the two divisions left on Peach Tree had a secure place of refuge in the *tête-de-pont* in the event I had moved out in that direction.

These preparations completed, he should on the morning of the morrow have ordered the divisions and cavalry, along this stream, to make demonstrations against the city whilst Thomas pushed forward in the direction of East Point—changing front forward on his left—and formed line of battle with his left flank resting as high up on Camp creek* as it would afford protection against its being turned, and his right extending to or across the West Point Railway; have instructed Schofield

* See Federals massed just below Camp creek, map, page 167.

and McPherson to move rapidly, as they had done upon Decatur and the Augusta road, to deploy on Thomas's right along the south bank of South river and east side of Shoal creek, with their right thrown back southeast of Decatur,* and to entrench the whole line.

Such would have been the position of the Federal Army within twenty-four hours after it left Peach Tree creek, and within ten days after its first crossing of the Chattahoochee, subsequent to the operations about Kennesaw Mountain, provided it had moved with the rapidity usual with the Confederate Armies. Even had forty-eight hours been required to perfect this movement, Sherman could have so manœuvred as to have held the main body of my troops on Peach Tree until he was willing I should become apprised of his real purpose. In other words, he could without difficulty have entrenched south of Atlanta, before I could have received the necessary information to warrant a change of position from the north to the south side of the city. Moreover, had I divined at an early instant his contemplated move, his position in rear of Peach Tree, and that of the two corps on Camp creek would—by demonstrations on the north and south sides of the city, with an Army double our own—have rendered it an easy matter to him to gain possession of Atlanta, in spite of every effort on my part.

General Sherman knew as well as I did, that every available man in the Confederacy had been sent either to General Lee, in Virginia, or to General Johnston, in the mountains; that, consequently, he had nothing to fear from the direction of Macon, and that one division would have sufficed to protect his rear, south of the city. When Grant marched round Pemberton at Vicksburg, and placed his rear in front of General Johnston, commanding an Army of twenty-five or thirty thousand men at Jackson, Mississippi, he executed successfully not only one of the boldest, but one of the grandest movements

* See line deployed from near East Point, map, page 167.

of the war. It will rank with one of the many similar moves of the immortal Jackson, and receive the tribute due to the talent and boldness which planned and achieved it. It was, however, fortunate for General Grant that a "Stonewall" was not at Jackson, Mississippi. No especial daring on the part of General Sherman would have been required to carry out the operations I have designated, since he had no enemy to fear in his rear. General Grant was reported, at this period, to have said that the Confederacy was but a shell.

As I have just remarked, I could not have received in time sufficiently reliable information to justify a change from the north to the south side of Atlanta, and to attack the Federals before they had thoroughly entrenched; it would have been equally impossible to assault later, with hope of success, his line, protected in front by works and abatis, on the left by Camp creek, and on the right by being thrown back and entrenched southeast of Decatur.

This position of the enemy would have necessitated the immediate abandonment of Atlanta or have shut up our Army in the pocket, or *cul de sac*, formed by the Chattahoochee river and Peach Tree creek, and finally have forced us to surrender. Had I attempted to extricate the Army, it would have been almost impossible to have pierced the enemy's works south, and utterly impossible, by reason of the proximity of the Federals, to have laid pontoons and crossed Peach Tree creek—as I would have done when Sherman was at the distance of Jonesboro', but from which I was hindered by the presence of the prisoners at Andersonville.

By reference to the map (page 167) it will be perceived that Sherman had simply to advance his right flank, in order to form a junction with the troops, near Decatur, and thus completely hem in our Army. This plan for the speedy capture of Atlanta could have been executed with an insignificant loss, as it would have been achieved mainly by manœuvre. In view of the impaired *morale* of the Army at the close of the Dalton-Atlanta campaign; the numerical inferiority of our forces; the fact, previously mentioned, that all available troops in the Confederacy, east of the Mississippi, had been sent either to the Army of Northern Virginia or to the Army of Tennessee, with the exception of small forces guarding the seaboard; in view of the proximity of the Chattahoochee river, which flows within five miles of Atlanta, along the foot of the general slope from the mountains of Georgia to the plains, forming with Peach Tree creek a complete *cul de sac*, in which Atlanta is situated; the advantages to be derived from Camp or any other creek in that vicinity, or from the deep ravines running to the river from the southwest side, behind which the Federal Army could have been rapidly deployed forward into line, with no enemy to fear from the rear; in which position Sherman could have readily supplied his Army from the nearest point on the railway north of the river, and have quietly awaited my surrender; in view, then, of the above enumerated sources of weakness and danger, I do not hesitate to challenge the military world to refute my assertions, not only as to the feasibility of the plan I have demonstrated, but also as to the untenability of Atlanta.

How long, I venture to inquire, is it probable that Sherman, after the capture of Jonesboro', would have tarried before occupying the identical position I have designated? The extraordinary haste I made to evacuate Atlanta, after the Federals gained possession of Jonesboro', on the Macon road, fifteen miles below the line from Camp creek to and along South river and Shoal creek, is proof of the great dread I entertained of a speedy occupation of this line. In lieu thereof, Sherman, during or immediately after the destruction of the Augusta road, threw Thomas across Peach Tree creek, into the *cul de sac* aforementioned, separated him from McPherson and Schofield, and subjected him to an assault by the main body of our Army, which should have resulted in the rout and capture of the greater portion of his Army. This move was, moreover, unnecessary, as it was impossible for him to invest Atlanta, approaching it from the north. He therefore consumed forty-six days in the achievement of results which might have been accomplished within a fortnight.

I have, from Dr. A. J. Foard, medical director of the Army, a statement of the total number of killed and wounded from July 18th to September 1st; in other words, from the day I assumed command to the evacuation of Atlanta. As I have already asserted, the number of men wounded in an Army which is standing its ground, and fighting, or is advancing, and driving the enemy, should not constitute an actual loss, and should only in part be comprised in the sum total of losses, since almost all the slightly wounded, proud of their scars, soon return to ranks. Therefore the only correct method of ascertaining the entire loss, during a siege or a campaign, is to deduct the number of effectives at the close of the siege, or campaign, from the number of effectives at the beginning of operations, after adding to the total strength the number of reinforcements, and deducting therefrom the number of troops permanently detached. This calculation should embrace the sick, and those who die from natural causes; losses under this head are, however, rarely of consequence during active operations. It is, as a generality, only when troops are lying in bivouac, or in quarters, that the ravages of disease are to be feared. Therefore, the fairness of this method can in no degree be affected by this minor fact.

Upon this broad and equitable basis, I herewith submit the official return of Colonel A. P. Mason, assistant adjutant general of the Army of Tennessee, showing its strength at different periods whilst under my command. As I desire to compare my strength and losses with Sherman's and Johnston's, I present, at the same time, Dr. Foard's official report of the killed and wounded; General Sherman's returns, showing his effective strength and estimate of losses; and the official statement of General Johnston's adjutant general, exhibiting the strength of the Army of Tennessee at different periods, during the campaign from Dalton to Atlanta.

"*Strength of the Army of Tennessee, on the 31st of July, 1864; 20th September, 1864; 6th November, 1864; and 10th December, 1864.*

July 31st, 1864.

	PRESENT.			ABSENT.	
	Effective.	Total.	Aggregate.	Total.	Aggregate.
Infantry..............	30,451	39,414	43,448	93,759	101,715
Cavalry..............	10,269	15,904	17,313	26,354	28,363
Artillery..............	3,775	4,610	4,840	6,317	6,606
Total Army...........	44,495	59,928	65,601	126,430	136,684

September 20th, 1864.

	PRESENT.			ABSENT.	
	Effective.	Total.	Aggregate.	Total.	Aggregate.
Infantry..............	27,094	36,301	39,962	81,824	89,030
Cavalry..............	10,543	15,978	17,416	27,005	29,215
Artillery..............	2,766	3,408	3,570	4,628	4,845
Total Army...........	40,403	55,687	60,948	113,457	123,090

November 6th, 1864.

	PRESENT.			ABSENT.	
	Effective.	Total.	Aggregate.	Total.	Aggregate.
Infantry..............	25,889	34,559	38,119	79,997	87,016
Cavalry..............	2,306	3,258	3,532	4,778	5,148
Artillery..............	2,405	2,913	3,068	4,018	4,203
Total Army...........	30,600	40,730	44,719	88,793	96,367

December 10th, 1864.

	PRESENT.			ABSENT.	
	Effective.	Total.	Aggregate.	Total.	Aggregate.
Infantry..............	18,342	27,222	29,826	71,329	77,631
Cavalry....	2,306	3,258	3,532	4,778	5,148
Artillery.............	2,405	2,913	3,068	4,018	4,203
*Total Army..........	23,053	33,393	36,426	80,125	86,982

"Respectfully submitted,

"A. P. MASON,

"Lieutenant Colonel, A. A. G."

"COLUMBUS, GEORGIA, *April 3d, 1866.*

"Consolidated summaries in the Armies of Tennessee and Mississippi during the campaign commencing May 7th, 1864, at Dalton, Georgia, and ending after the engagement with the enemy at Jonesboro' and the evacuation of Atlanta, furnished for the information of General J. E. Johnston.

"*Consolidated summary of casualties of the Armies of Tennessee and Mississippi in the series of engagements around and from Dalton, Georgia, to the Etowah river, for the period commencing May the 7th, and ending May 20th, 1864:*

CORPS.	Killed.	Wounded.	Total.
Hardee's........................	119	859	978
Hood's..........................	283	1,564	1,847
Polk's Army, Mississippi......	42	405	447
	444	2,828	3,372

*It should be remembered that, in all estimates of the strength of Armies, the number of *effectives* is alone to be considered; therefore, the first column, in the foregoing return, is that to which reference should be made. Also, that of the forty thousand four hundred and three (40,403) effectives reported present for duty on the 20th September, forty-five hundred (4500) cavalry were absent with Wheeler, in Tennessee. This latter circumstance accounts for my statement, subsequently, that we had thirty-five thousand (35,000) effectives during the

"*Consolidated summary of casualties of the Armies of Tennessee and Mississippi in the series of engagements around New Hope Church, near Marietta, Georgia:*

Corps.	Killed.	Wounded.	Total.
Hardee's	173	1,048	1,221
Hood's	103	679	782
Polk's Army, Mississippi	33	194	227
	309	1,921	2,230

"*Consolidated summary of casualties of the Armies of Tennessee and Mississippi in the series of engagements around Marietta, Georgia, from June 4th to July 4th, 1864:*

Corps.	Killed.	Wounded.	Total.
Hardee's	200	1,433	1,633
Hood's	140	1,121	1,261
Polk's Army, Mississippi	128	926	1,054
	468	3,480	3,948

"*Consolidation of the above three Reports is as follows:*

	Killed.	Wounded.	Total.
Dalton to Etowah river	444	2,828	3,272
New Hope Church	309	1,921	2,230
Around Marietta	468	3,480	3,948
	1,221	8,229	9,450

campaign to the Alabama line. It should, in addition, be observed that Wheeler's cavalry, ten thousand five hundred and forty-three (10,543) in number, as borne upon Colonel Mason's return, on the 20th September, was left in Georgia when we crossed the Tennessee, and was replaced by Forrest's cavalry, numbering altogether two thousand three hundred and six (2306) effectives. This large detachment will account for the reduction in the strength of our Army, at Palmetto and Florence, as will be seen later in my narrative of the campaign to the Alabama line, and thereafter into Tennessee.

"*Consolidated summary of casualties of the Army of Tennessee (Army of Mississippi being merged into it) in the series of engagements around Atlanta, Georgia, commencing July 4th, and ending July 31st, 1864:*

Corps.	Killed.	Wounded.	Total.
Hardee's	523	2,774	3,297
Lee's	351	2,408	2,759
Stewart's	436	2,141	2,577
Wheeler's Cavalry	29	156	185
Engineers	2	21	23
	1,341	7,500	8,841

"*Consolidated summary of casualties in Army of Tennessee in engagements around Atlanta and Jonesboro', from August 1st to September 1st, 1864:*

Corps.	Killed.	Wounded.	Total.
Hardee's	141	1,018	1,159
Lee's	248	1,631	1,879
Stewart's	93	574	667
	482	3,223	3,705

"*Consolidation of which two Reports is as follows:*

	Killed.	Wounded.	Total.
Around Atlanta, July 4th to July 31st, 1864	1,341	7,500	8,841
Atlanta and Jonesboro', Aug. 1st to Sept. 1st, 1864	482	3,223	3,705
	1,823	10,723	12,546

"I certify that the above reports are from the returns made to my office, and are in my opinion correct.

"(Signed) A. J. FOARD."

"Medical Director, late Army of Tennessee."

"NOTE.—The Atlanta-Dalton campaign began on May 7th, and ended on the 1st of September, 1864, and the above reports are exact copies of those made to the Commanding General during its progress, and in the order in which they here appear.

"General Johnston commanded from the commencement of the campaign until the 18th of July, when he was relieved from duty, and General Hood assigned to the command of the Army. Hence the casualties of battle which occurred in the Army between the 4th and the 18th of July belong to the period of General Johnston's command, and are as follows: Killed, sixty-seven (67); wounded, four hundred and fifty-five (455); total, five hundred and twenty-two (522). These figures, added to the total of casualties as reported up to July 4th, viz., killed, twelve hundred and twenty-one (1221), wounded, eight thousand two hundred and twenty-nine (8229). total, nine thousand four hundred and fifty (9450), gives the entire losses (killed and wounded) in battle for the whole Army, while under the command of General Johnston, as follows, viz: killed, twelve hundred and eighty-eight (1288); wounded, eight thousand six hundred and eighty-four (8684); total, nine thousand nine hundred and seventy-two (9972). A deduction of the same, viz., killed, sixty-seven (67), wounded, four hundred and fifty-five (455), total, five hundred and twenty-two (522), from the total of casualties reported from July 4th to September 1st, viz., killed, eighteen hundred and twenty-three (1823), wounded, ten thousand seven hundred and twenty-three (10,723), total, twelve thousand five hundred and forty-six (12,546), gives of killed seventeen hundred and fifty-six (1756), wounded, ten thousand two hundred and sixty-seven (10,267); total, twelve thousand and twenty-three (12,023), as the entire losses in killed and wounded during that period of the campaign when the Army was commanded by General Hood, viz., from July the 18th to September 1st, 1864, when it ended, and the Army was then prepared for the campaign into Tennessee.

"(Signed) A. J. FOARD,

"Medical Director late Army of Tennessee."

"*On recapitulating the entire losses of each Army during the entire campaign, from May to September, inclusive, we have, in the Union Army, as per table appended:* *

Killed	4,423
Wounded	22,822
Missing	4,442
Aggregate loss	31,687."

* Sherman's Memoirs, page 132, vol. II.

"HEADQUARTERS MILITARY DIVISION OF THE MISSISSIPPI,
"IN THE FIELD, ATLANTA, GEORGIA, *September 15th, 1864.*

"*Prisoners and deserters taken by 'Army in the Field,' Military Division of the Mississippi, during May, June, July, and August, 1864:**

COMMANDS.	PRISONERS.		DESERTERS.		
	Officers.	Men.	Officers.	Men.	Aggregate.
Army of the Cumberland,	121	3,838	21	1,543	5,523
" " Tennessee ..	133	2,591	5	576	3,305
" " Ohio	16	781	1	292	1,090
Total............	270	7,210	27	2,411	9,918

"SHERMAN'S FORCES.†

"*Recapitulation—Atlanta Campaign.*

Arm.	June 1.	July 1.	August 1.	Sept 1.
Infantry................	94,310	88,066	75,659	67,674
Cavalry................	12,908	12,039	10,517	9,394
Artillery...............	5,601	5,945	5,499	4,690
Aggregate.............	112,819	106,050	91,675	81,758

"NEAR GREENSBORO," NORTH CAROLINA,
"*May 1, 1865.*

"1. *The 'effective strength' of the Army of Tennessee, as shown by the tri-monthly return of the 1st of May, 1864, was:*‡

Infantry.............................37,652 } 40,464
Artillery............................ 2,812 }
Cavalry.............................. 2,392

"This was the entire strength of the Army 'at and near Dalton' at that date.

"2. The movement from Dalton began on the 12th May. On that day Loring's Division, Army of Mississippi, and Cantry's Division,

* Sherman's Memoirs, vol. II, page 134.

† Sherman's Memoirs, vol. II, page 136.

‡ Johnston's Narrative, pages 574, 575.

joined at Resaca, with about eight thousand (8000) effectives. French's Division, same Army, joined near Kingston several days later (about four thousand (4000) effectives). Quarles's brigade from Mobile (about twenty-two hundred (2200) effectives) joined at New Hope Church on the 26th. The cavalry of the Mississippi Army, which joined near Adairsville, was estimated at three thousand nine hundred (3900) effectives; and Martin's Cavalry Division, which joined near Resaca, at three thousand five hundred (3500). These were the only reinforcements received while General Johnston had command of the Army.

"3. There was no return (filed) of the Army made after May 1st, until June 10th. The return of June 10th gave, as effectives:

Infantry	44,860	48,732
Artillery	3,872	
Cavalry	10,516	

"4. The next return was made on the 1st of July.

Effectives:	Infantry	39,197	42,666
	Artillery	3,469	
	Cavalry	10,023	

"On the 3d of July, at Vining's Station, the Fifth and Forty-seventh Georgia Regiments (about six hundred (600) effectives) left the Army for Savannah, under Brigadier General J. K. Jackson.

"5. The next and last return made under General Johnston was on the 10th of July.

Effectives:	Infantry	36,901	40,656
	Artillery	3,755	
	Cavalry	9,971	(Exclusive of escorts serving with infantry.)

"This was the estimated force turned over by General Johnston to General Hood.

"6. The report was made under General Johnston, and signed by General Hood. On the 18th of July the command was turned over to General Hood. The first return thereafter was that of August 1st, after the engagements of Peach Tree creek, on the 21st, and around Atlanta, on the 22d and 28th July.

"7. The foregoing figures are taken from the official records kept by me as assistant adjutant general of the Army.

"(Signed) KINLOCH FALCONER,

"Assistant Adjutant General."

I here reiterate that it is impossible General Johnston should have turned over to me fifty thousand six hundred and twenty-seven (50,627) effectives on the 18th of July (as shown in Colonel Falconer's report), for the reason that he had this number in full on the 10th of that month. When, according to this same report, we suffered a loss, over and above the killed and wounded, of four thousand and seventy-three (4073) men who abandoned their colors, and went either to their homes or to the enemy just prior to the retreat across the Chattahoochee river, it is not reasonable to assume that no desertions occurred from the 10th of July—the date of his last return—to the 18th, when a change of commanders took place in the face of the enemy, and under extraordinary circumstances. The supposition that many deserted during this interval is but just and natural. I am, therefore, confident that I am over-liberal in the estimate given—forty-eight thousand seven hundred and fifty (48,750) effectives—in my official report of the effective strength of the Army of Tennessee, when I assumed command. However, I will, in this instance grant, for the sake of argument, that my force on the 18th of July was fifty thousand six hundred and twenty-seven (50,627) effectives.

On the 20th of September, when stragglers had been gathered up, the effective strength of the Confederate Army, according to Colonel Mason's report, was forty thousand four hundred and three (40,403). This number, subtracted from fifty thousand six hundred and twenty-seven (50,627)—less thirty-one hundred (3100) permanently detached to Macon and Mobile, about the beginning of the siege—shows a loss of seven thousand one hundred and twenty-four (7124), to which should be added two thousand prisoners returned to the ranks by exchange, soon after the fall of Atlanta, and before Colonel Mason made up his return on the 20th of September. These prisoners were overlooked by myself and my chief of staff at the time I made my official report, and increase the total loss, from all causes, to nine thousand one hundred and twenty-four (9124).

Whilst Dr. Foard's report of the killed and wounded is correct, the above estimate is beyond doubt equally accurate, since I received no reinforcements, during the siege, which were not sent back soon after their arrival, with the exception of about two hundred and fifty men of Gholsen's brigade (which small force I have not taken into account), as the following letter from General Shoupe will indicate:

"RICHMOND, *March 10th, 1865.*

"GENERAL HOOD:—You ask to what extent your Army was strengthened at Atlanta by the return of detailed men, and by dismounted cavalry ordered to you by General Bragg. I have the honor to state that so far as the detailed men are concerned, it was found necessary to return them to the arsenals and shops in rear, and that they were, as I believe, all so returned before the evacuation of Atlanta. Roddy's cavalry, upon the very day it reached Atlanta, was ordered back to Alabama. Gholsen's brigade remained at Atlanta until its evacuation. It was, however, very small—not numbering more than two hundred and fifty (250) men, and was in most miserable condition. So that the reinforcements, in truth, amounted to nothing.

"I have the honor to be very respectfully, etc.,

"F. A. SHOUPE,

"Brigadier General and Chief of Staff at Atlanta."

Although the number of killed and wounded in the Army of Tennessee proper, during the siege, amounted to twelve thousand and twenty-three (12,023), the actual loss was nine thousand one hundred and twenty-four (9124); thus proving that near three thousand wounded returned to the ranks.

I shall now sum up the loss of the enemy during that same period.

* General Sherman reports his loss in killed, wounded, and missing, around Atlanta during July, August and September, to have been fifteen thousand and thirty-three (15,033). His actual loss during the siege must assuredly have been in excess of this number. In accordance with his recapitulation,†

---

* Sherman's Memoirs, vol. II, page 133.

† Sherman's Memoirs, vol. II, page 136.

he had on the 1st of July an Army of one hundred and six thousand and seventy (106,070); on the 1st of August, ninety-one thousand six hundred and seventy-five (91,675); and on the 1st of September, eighty-one thousand seven hundred and fifty-eight (81,758), demonstrating an actual loss of twenty-four thousand three hundred and twelve (24,312) men within two months. This number, less the troops discharged or permanently detached, must be the real loss he sustained. I have not been able to glean from his statements the decrease of his Army from this latter source. I find, however, the following recorded in Shoupe's Diary on the 17th of August:

"Enemy's pickets called to ours, and stated that a Kentucky Division, twenty-two hundred (2200) strong, was going out of service, and that neither Old Abe nor Uncle Jeff would get them in service again."

Taking his own statements as a basis of calculation, and assuming the correctness of the report by the picket relative to the discharge of twenty-two hundred (2200) Kentuckians thirteen days prior to the fall of Atlanta, his actual losses (provided he did not during the siege receive reinforcements, of which I can find no mention in his Memoirs), prove to have been twenty-four thousand three hundred and twelve (24,312), plus nineteen hundred and two (1902) killed and wounded early in September, minus twenty-two hundred (2200) discharged; showing an actual loss of twenty-four thousand and fourteen (24,014) effectives against my loss of nine thousand one hundred and twenty-four (9124), although every aggressive movement of importance was initiated by the Confederates.

On the other hand, and according to my opponent's statement,* General Sherman had, after Blair's Corps joined him near Rome, a force of one hundred and twelve thousand eight hundred and nineteen (112,819) effectives to oppose General Johnston; and at the close of his victorious march from Dalton to Atlanta, one hundred and six thousand and seventy (106,070) effectives, which subtracted from the total number

* Sherman's Memoirs, vol. II, page 136.

one hundred and twelve thousand eight hundred and nineteen (112,819) in the field, at the beginning of the campaign, demonstrates an actual loss of only six thousand seven hundred and forty-nine (6749) against General Johnston's loss of twenty-five thousand (25,000) men.

This comparison of losses under opposite modes of handling troops, evinces the truth of the principle for which I contend: that losses are always comparatively small in an Army which drives before it the enemy day after day, as in the instance of the Federal Army during the Dalton-Atlanta campaign; or in an Army which holds its ground, as in the instance of the siege of Atlanta when the Federal loss was greatly in excess of our own, by reason of the enthusiasm and self-reliance of the Northern troops having, in the sudden check given to their sweeping career of victory, been somewhat counteracted by depression, consequent desertion, and the tardy return of absentees.

CHAPTER XIV.

CORRESPONDENCE WITH SHERMAN—CITATIONS ON THE RULES OF WAR.

ABOUT the time I exchanged with General Sherman the two thousand (2000) prisoners above mentioned, the following correspondence passed between us, in relation to his treatment of the non-combatants of Atlanta:

"HEADQUARTERS MILITARY DIVISION OF THE MISSISSIPPI,
"IN THE FIELD, ATLANTA, GEORGIA, *September 7th, 1864.*

"GENERAL HOOD, *Commanding Confederate Army.*

"GENERAL:—I have deemed it to the interest of the United States that the citizens now residing in Atlanta should remove, those who prefer it to go South, and the rest North. For the latter I can provide food and transportation to points of their election in Tennessee, Kentucky, or further North. For the former I can provide transportation by cars as far as Rough and Ready, and also wagons; but, that their removal may be made with as little discomfort as possible, it will be necessary for you to help the families from Rough and Ready to the cars at Lovejoy's. If you consent, I will undertake to remove all the families in Atlanta who prefer to go South to Rough and Ready, with all their moveable effects, viz., clothing, trunks, reasonable furniture, bedding, etc., with their servants, white and black, with the proviso that no force shall be used toward the blacks, one way or another. If they want to go with their masters or mistresses, they may do so; otherwise they will be sent away unless they be men, when they may be employed by our quarter-master.

Atlanta is no place for families or non-combatants, and I have no desire to send them North if you will assist in conveying them South. If this proposition meets your views, I will consent to a truce in the neighborhood of Rough and Ready, stipulating that any wagons, horses, animals, or persons sent there for the purposes herein stated, shall in no manner be harmed or molested; you in your turn agreeing that any cars, wagons, or carriages, persons or animals sent to the same point, shall not be interfered with. Each of us might send a guard of, say one hundred (100) men, to maintain order; and limit the truce to, say, two days after a certain time appointed.

"I have authorized the Mayor to choose two citizens to convey to you this letter, with such documents as the Mayor may forward in explanation and shall await your reply.

"I have the honor to be your obedient servant,

"W. T. SHERMAN,

"Major General Commanding."

"HEADQUARTERS ARMY OF TENNESSEE,
"OFFICE CHIEF OF STAFF, *September 9th, 1864.*

"MAJOR GENERAL W. T. SHERMAN, *Commanding United States Forces in Georgia.*

"GENERAL:—Your letter of yesterday's date, borne by James M. Ball and James R. Crew, citizens of Atlanta, is received. You say therein, 'I deem it to be to the interest of the United States that the citizens now residing in Atlanta should remove,' etc.

"I do not consider that I have any alternative in this matter. I therefore accept your proposition to declare a truce of two days, or such time as may be necessary to accomplish the purpose mentioned, and shall render all assistance in my power to expedite the transportation of citizens in this direction. I suggest that a staff officer be appointed by you to superintend the removal from the city to Rough and Ready, while I appoint a like officer to control their removal further South; that a guard of one hundred men be sent by either party as you propose, to maintain order at the place, and that the removal begin on Monday next.

"And now, sir, permit me to say that the unprecedented measure you propose transcends, in studied and ingenious cruelty, all acts ever before brought to my attention in the dark history of war.

"In the name of God and humanity, I protest, believing that you will find that you are expelling from their homes and firesides the wives and children of a brave people.

"I am, General, very respectfully, your obedient servant,

"J. B. HOOD, *General.*"

"HEADQUARTERS MILITARY DIVISION OF THE MISSISSIPPI,
"IN THE FIELD, ATLANTA, GEORGIA, *September 10th, 1864.*

"GENERAL J. B. HOOD, *Commanding Army of Tennessee, Confederate Army.*

"GENERAL:—I have the honor to acknowledge the receipt of your letter of this date, at the hands of Messrs. Ball and Crew, consenting to the arrangements I had proposed to facilitate the removal South of the people of Atlanta, who prefer to go in that direction. I enclose you a copy of my orders, which will, I am satisfied, accomplish my purpose perfectly.

"You style the measures proposed 'unprecedented,' and appeal to the dark history of war for a parallel, as an act of 'studied and ingenious cruelty.' It is not unprecedented; for General Johnston himself very wisely and properly removed the families all the way from Dalton down, and I see no reason why Atlanta should be excepted. Nor is it necessary to appeal to the dark history of war, when recent and modern examples are so handy. You yourself burned dwelling houses along your parapet, and I have seen to-day fifty houses that you have rendered uninhabitable because they stood in the way of your forts and men. You defended Atlanta on a line so close to town that every cannon shot and many musket shots from our line of investment, that overshot their mark, went into the habitations of women and children. General Hardee did the same at Jonesboro', and General Johnston did the same, last summer, at Jackson, Mississippi. I have not accused you of heartless cruelty, but merely instance these cases of very recent occurrence, and could go on and enumerate hundreds of others, and challenge any fair man to judge which of us has the heart of pity for the families of a 'brave people.'

"I say it is kindness to these families of Atlanta to remove them now, at once, from scenes that women and children should not be exposed to, and the 'brave people' should scorn to commit their wives and children to the rude barbarians who thus, as you say, violate the laws of war, as illustrated in the pages of its dark history.

"In the name of common sense, I ask you not to appeal to a just God in such a sacrilegious manner. You who, in the midst of peace and prosperity, have plunged a nation into war—dark and cruel war—who dared and badgered us to battle, insulted our flag, seized our arsenals and forts that were left in the honorable custody of peaceful ordnance sergeants, seized and made 'prisoners of war' the very garrisons sent to protect your people against negroes and Indians, long before any overt act was committed by the (to you) hated Lincoln Government; tried to force Kentucky and Missouri into rebellion, spite of themselves; falsified the vote of Louisiana; turned loose your privateers to plunder unarmed ships; expelled Union families by the thousands, burned their houses, and declared, by an act of your Congress, the confiscation of all debts due Northern men for goods had and received! Talk thus to the marines, but not to me, who have seen these things, and who will this day make as much sacrifice for the peace and honor of the South as the best born Southerner among you! If we must be enemies, let us be men, and fight it out as we propose to do and not deal in such hypocritical appeals to God and humanity. God will judge us in due time, and he will pronounce whether it be more humane to fight with a town full of women and the families of a brave people at our back, or to remove them in time to places of safety among their own friends and people.

"I am, very respectfully, your obedient servant,

"W. T. SHERMAN,

"Major General Commanding."

"HEADQUARTERS ARMY OF TENNESSEE,
"*September 12th, 1864.*

"MAJOR GENERAL W. T. SHERMAN, *Commanding Military Division of the Mississippi.*

"GENERAL:—I have the honor to acknowledge the receipt of your letter of the 9th inst., with its inclosure in reference to the women, children, and others, whom you have thought proper to expel from their homes in the city of Atlanta. Had you seen proper to let the matter rest there, I would gladly have allowed your letter to close this correspondence, and, without your expressing it in words, would have been willing to believe that, while 'the interests of the United States,' in your opinion, compelled you to an act of barbarous cruelty, you regretted the necessity, and we would have dropped the subject; but you have chosen to indulge in statements which I feel compelled to notice, at least so far as to signify my dissent, and not allow silence in regard to them to be construed as acquiescence.

"I see nothing in your communication which induces me to modify the language of condemnation with which I characterized your order. It but strengthens me in the opinion that it stands 'pre-eminent in the dark history of war for studied and ingenious cruelty.' Your original order was stripped of all pretences; you announced the edict for the sole reason that it was to 'the interest of the United States.' This alone you offered to us and the civilized world as an all-sufficient reason for disregarding the laws of God and man. 'You say that General Johnston himself very wisely and properly removed the families all the way from Dalton down.' It is due to that gallant soldier and gentleman to say that no act of his distinguished career gives the least color to your unfounded aspersions upon his conduct. He depopulated no villages, nor towns, nor cities, either friendly or hostile. He offered and extended friendly aid to his unfortunate fellow-citizens who desired to flee from your fraternal embraces. You are equally unfortunate in your attempt to find a justification for this act of cruelty, either in the defence of Jonesboro', by General Hardee, or of Atlanta, by myself. General Hardee defended his position in front of Jonesboro' at the expense of injury to the houses; an ordinary, proper, and justifiable act of war. I defended Atlanta at the same risk and cost. If there was any fault in either case, it was your own, in not giving notice, especially in the case of Atlanta, of your purpose to shell the town, which is usual in war among civilized nations. No inhabitant was expelled from his home and fireside by the orders of General Hardee or myself, and therefore your recent order can find no support from the conduct of either of us. I feel no other emotion other than pain in reading that portion of your letter which attempts to justify your shelling Atlanta, without notice, under pretence that I defended Atlanta upon a line so close to town that every cannon shot, and many musket balls from your line of investment, that overshot their mark, went into the habitations of women and children. I made no complaint of your firing into Atlanta in any way you thought proper. I make none now, but there are a hundred thousand witnesses that you fired into the habitations of women and children for weeks, firing far above and miles beyond my line of defence. I have too good an opinion, founded both upon observation and experience, of the skill of your artillerists, to credit the insinuation that they for several weeks unintentionally fired too high for my modest field-works, and slaughtered women and children by accident and want of skill.

"The residue of your letter is rather discussion. It opens a wide field for the discussion of questions which I do not feel are committed to me. I am only a General of one of the Armies of the Confederate States, charged with military operations in the field, under the direction of my superior officers, and I am not called upon to discuss with you the causes of the present war, or the political questions which led to or resulted from it. These grave and important questions have been committed to far abler hands than mine, and I shall only refer to them so far as to repel any unjust conclusion which might be drawn from my silence. You charge my country with 'daring and badgering you to battle.' The truth is, we sent commissioners to you, respectfully offering a peaceful separation, before the first gun was fired on either side. You say we insulted your flag. The truth is, we fired upon it, and those who fought under it, when you came to our doors upon the mission of subjugation. You say we seized upon your forts and arsenals, and made prisoners of the garrisons sent to protect us against Indians and negroes. The truth is, we, by force of arms, drove out insolent intruders and took possession of our own forts and arsenals, to resist your claims to do minion over masters, slaves, and Indians, all of whom are to this day, with a unanimity unexampled in the history of the world, warring against your attempts to become their masters. You say that we tried to force Kentucky and Missouri into rebellion in spite of themselves. The truth is, my Government, from the beginning of this struggle to this hour, has again and again offered, before the whole world, to leave it to the unbiassed will of these States, and all others, to determine for themselves whether they will cast their destiny with your Government or ours; and your Government has resisted this fundamental principle of free institutions with the bayonet, and labors daily, by force and fraud, to fasten its hateful tyranny upon the unfortunate freemen of these States. You say we falsified the vote of Louisiana. The truth is, Louisiana not only separated herself from your Government by nearly a unanimous vote of her people, but has vindicated the act upon every battle-field from Gettysburg to the Sabine, and has exhibited an heroic devotion to her decision, which challenges the admiration and respect of every man capable of feeling sympathy for the oppressed or admiration for heroic valor. You say that we turned loose pirates to plunder your unarmed ships. The truth is, when you robbed us of our part of the Navy, we built and bought a few vessels, hoisted the flag of our country, and swept the seas, in defiance of your Navy, around the whole circumference of the globe. You say we have expelled Union families by thousands. The truth is, not a single family has been expelled from the Confederate States, that I am aware of; but, on the contrary, the moderation of our Government towards traitors has been a fruitful theme of denunciation by its enemies and well meaning friends of our cause. You say my Government, by acts of Congress, has confiscated 'all debts due Northern men for goods sold and delivered.' The truth is, our Congress gave due and ample time to your merchants and traders to depart from our shores with their ships, goods, and effects, and only sequestrated the property of our enemies in retaliation for their acts—declaring us traitors, and confiscating our property wherever their power extended, either in their country or our own. Such are your accusations, and such are the facts known of all men to be true.

"You order into exile the whole population of a city; drive men, women, and children from their homes at the point of the bayonet, under the plea that it is to the interest of your Government, and on the claim that it is an act of 'kindness to these families of Atlanta.' Butler only banished from New Orleans the registered enemies of his Government, and acknowledged that he did it as a punishment. You issue a sweeping edict, covering all the inhabitants of a city, and add insult to the injury heaped upon the defenceless by assuming that you have done them a kindness. This you follow by the assertion that you 'will make as much sacrifice for the peace and honor of the South as the best born Southerner.' And, because I characterize what you call a kindness as being real cruelty, you presume to sit in judgment between me and my God; and you decide that my earnest prayer to the Almighty Father to save our women and children from what you call kindness, is a 'sacrilegious, hypocritical appeal.'

"You came into our country with your Army, avowedly for the purpose of subjugating free white men, women, and children, and not only intend to rule over them, but you make negroes your allies, and desire to place over us an inferior race, which we have raised from barbarism to its present position, which is the highest ever attained by that race, in any country, in all time. I must, therefore, decline to accept your statements in reference to your kindness toward the people of Atlanta, and your willingness to sacrifice everything for the peace and honor of the South, and refuse to be governed by your decision in regard to matters between myself, my country, and my God.

"You say, 'let us fight it out like men.' To this my reply is—for myself, and I believe for all the true men, ay, and women and children, in my country—we will fight you to the death! Better die a thousand deaths than submit to live under you or your Government and your negro allies!

"Having answered the points forced upon me by your letter of the 9th of September, I close this correspondence with you; and, notwithstanding your comments upon my appeal to God in the cause of humanity, I again humbly and reverently invoke his Almighty aid in defence of justice and right.

"Respectfully, your obedient servant,

"J. B. HOOD, *General.*"

"HEADQUARTERS MILITARY DIVISION OF THE MISSISSIPPI,
"ATLANTA, GA., *September, 14th, 1864.*

"GENERAL J. B. HOOD, *Commanding Army of Tennessee.*

"GENERAL:—Yours of September 12th is received, and has been carefully perused. I agree with you that this discussion by two soldiers is out of place, and profitless; but you must admit that you began the controversy by characterizing an official act of mine in unfair and improper terms. I reiterate my former answer, and to the only new matter contained in your rejoinder add: We have no 'negro allies' in this Army; not a single negro soldier left Chattanooga with this Army, or is with it now. There are a few guarding Chattanooga, which General Stedman sent at one time to drive Wheeler out of Dalton.

"I was not bound by the laws of war to give notice of the shelling of Atlanta, a 'fortified town, with magazines, arsenals, foundries, and public stores;' you were bound to take notice. See the books.

"This is the conclusion of our correspondence, which I did not begin, and terminate with satisfaction.

"I am, with respect, your obedient servant,

"W. T. SHERMAN,

"Major General Commanding."

I preferred here to close the discussion, and, therefore, made no reply to his last communication inviting me to "see the books."

I will at present, however, consider this subject, and cite a few authorities upon the above disputed points, in order to show that General Sherman's conduct, in this instance, was in violation of the laws which should govern nations in time of war.

Atlanta could not properly be designated a regularly fortified city. It was simply protected by temporary breastworks, of the same character as those used by Johnston and Sherman, during the preceding campaign. The fortifications consisted of a ditch, with a log to act as protection to the heads of the men whilst firing, and of brushwood, when it could be obtained, thrown out in front as an obstruction to a rapid advance of the enemy. A large portion of the line, which passed through open fields, was devoid of this latter safeguard. Moreover, only a few of the heavy guns and batteries were covered by embankments with embrasures.

Fortifications, it is well known, are divided into two classes: temporary, and permanent. Those I have described, around Atlanta, come under the head of the first class. The latter are constructed of the best material, iron, and stone, with parapet, deep and wide ditch and glacis, similar to the fortifications on Governor's Island, and those of Fortress Monroe. In the construction of permanent works, every exertion is made to render them as strong and durable as possible.

It might be supposed, from General Sherman's Memoirs, that Atlanta was not only a thoroughly fortified town, but was provisioned to endure a siege of a year or more, after all communication was cut off; that it possessed arsenals and machine shops as extensive as those in Richmond and Macon—an illusion created, probably, by a dilapidated foundry, near the Augusta road, which had been in use prior to the war. General Sherman, therefore, cannot assert, in order to justify certain acts, that Atlanta was a regularly fortified town. And whereas I marched out at night, allowing him the following day to enter the city, unopposed, as he himself acknowledges, and whereas no provocation was given by the authorities, civil or military, he can in no manner claim that extreme war measures were a necessity.

It has been argued that Wellington sanctioned extreme measures against the Basques, at the time he was opposed to Marshal Soult, at Bayonne, in 1814. Wellington perceived that, by pillage and cruel treatment, his Spanish allies, under Mina and Morilla, were arousing the Basques to arms, and at once ordered the Spanish troops to abstain from such odious conduct. He was, unfortunately, too late in his discovery; the appetite for plunder had become so inordinate that his proclamation was disregarded by his allies, and he was subsequently forced to threaten extreme measures, in order to check the partisan warfare which initiated the cruelties and horrors he deplored. This is the unquestionable interpretation of the subjoined passage: "A sullen obedience followed, but the plundering system was soon renewed, and this, with the mischief already done, was enough to arouse the inhabitants of *Bedary*, as well as those of the Val-de-Baigorre, into action. They commenced and continued a partisan warfare until Lord Wellington, incensed by their activity, issued a proclamation calling upon them to take arms openly, and join Soult, or stay peaceably at home, declaring that he would otherwise burn their villages and hang all the inhabitants." *

* Peninsular War, B. XXIII, chap. 3.

The inhabitants of Atlanta gave no such cause for action on the part of General Sherman, nor was the safety of the Federal Army in any manner involved. Nevertheless he ordered women and children, the infirm and the sick, in fact the entire population to go either North or South.

The subjoined appeal of the Mayor and Councilmen of Atlanta was powerless to alter the determination of the Federal commander:

"ATLANTA, GEORGIA, *September 11th, 1864.*

"MAJOR GENERAL W. T. SHERMAN.

"SIR:—We, the undersigned, Mayor and two of the Council for the city of Atlanta, for the time being the only legal organ of the people of the said city, to express their wants and wishes, ask leave most earnestly but respectfully to petition you to reconsider the order requiring them to leave Atlanta.

"At first view, it struck us that the measure would involve extraordinary hardship and loss, but since we have seen the practical execution of it, so far as it has progressed, and the individual condition of the people, and heard their statements as to the inconveniences, loss, and suffering attending it, we are satisfied that the amount of it will involve in the aggregate consequences appalling and heart-rending.

"Many poor women are in advanced state of pregnancy, others now having young children, and whose husbands, for the greater part, are either in the Army, prisoners, or dead.

"Some say: 'I have such a one sick at my house; who will wait on them when I am gone?' Others say: 'What are we to do? We have no house to go to, and no means to buy, build, or rent any; no parents, relatives, or friends to go to.' Another says: 'I will try and take this or that article of property, but such and such things I must leave behind, though I need them much.' We reply to them: 'General Sherman will carry your property to Rough and Ready, and General Hood will take it thence on.' And they will reply to that: 'But I want to leave the railroad at such a place, and cannot get conveyance from there on.'

"We only refer to a few facts, to try to illustrate in part how this measure will operate in practice. As you advanced, the people north of this fell back, and before your arrival here, a large portion of the people had retired South; so that the country south of this is already crowded, and without houses enough to accommodate the people, and we are informed that many are now staying in churches and other out-buildings.

"This being so, how is it possible for the people still here (mostly women and children) to find any shelter? And how can they live through the Winter in the woods—no shelter or subsistence, in the midst of strangers who know them not, and without the power to assist them much, if they were willing to do so.

"This is but a feeble picture of the consequences of this measure. You know the woe, the horrors, and the sufferings cannot be described by words; imagination can only conceive it, and we ask you to take these things into consideration.

"We know your mind and time are constantly occupied with the duties of your command, which almost deters us from asking your attention to this matter, but thought it might be that you had not considered this subject in all of its awful consequences, and that on more reflection you, we hope, would not make this people an exception to all mankind; for we know of no such instance ever having occurred—surely never in the United States—and what has this *helpless* people done, that they should be driven from their homes, to wander strangers, and outcasts, and exiles, and to subsist on charity?

"We do not know as yet the number of people still here; of those who are here, we are satisfied a respectable number, if allowed to remain at home, could subsist for several months without assistance, and a respectable number for a much longer time, and who might not need assistance at any time.

"In conclusion, we most earnestly and solemnly petition you to reconsider this order, or modify it, and suffer this unfortunate people to remain at home, and enjoy what little means they have.

"Respectfully submitted,

"JAMES M. CALHOUN, *Mayor.*
"E. E. RAWSON, *Councilman.*
"S. C. WELLS, *Councilman.*"

I shall now cite a few authorities upon the rights of war, to ascertain in how far the course pursued toward the inhabitants of Atlanta is in accordance with those laws which are now universally recognized.

Halleck, Vattel, and Grotius establish the following rules:*

"* * * It is a just remark made by some theologians, that all *Christian* princes and rulers who wish to be found *such* in the sight of God, as well as that of men, will deem it a duty to interpose their authority to

*Grotius, B. III, chap. 12, sec. 8. (The italics are the author's.)

prevent or suppress all *unnecessary* violence in the taking of terms, for acts of rigor can never be carried to an extreme without involving great numbers of the innocent in ruin; and practices of that kind, beside being no way conducive to the termination of war, are totally repugnant to every principle of Christianity and justice."

"Women, children, feeble old men, and sick persons, come under the description of enemies; and we have certain rights over them, inasmuch as they belong to the nation with whom we are at war, and as, between nation and nation, all rights and pretensions affect the body of society, together with all its members. But these are enemies who make no resistance; and consequently we have no right to maltreat their persons, or use any violence against them. * * * This is so plain a maxim of justice and humanity that at present every nation, in the least degree civilized, acquiesces in it."* * * *

"Since women and children are subjects of the State, and members of the Nation, they are to be ranked in the class of enemies. But it does not thence follow that we are justifiable in treating them like men who bear arms, or are capable of bearing them." †

"At present war is carried on by regular troops; the people, the peasants, the citizens, take no part in it, and generally have nothing to fear from the sword of the enemy. Provided the inhabitants submit to him who is master of the country, pay the contributions imposed, and refrain from all hostilities, they live in as perfect safety as if they were friends; they continue in possession of what belongs to them; the country people come freely to the camp to sell their provisions, and are protected as far as possible from the calamities of war." ‡

"Since the object of a just war is to repress injustice and violence, and forcibly to compel him who is deaf to the voice of justice, we have a right to put in practice, against the enemy, every measure that is necessary in order to weaken him and disable him from resisting us and supporting his injustice; and we may choose such methods as are the most efficacious and best calculated to attain the end in view, provided they be not of an odious kind, nor unjustifiable in themselves, and prohibited by the laws of nature." §

"The lawfulness of the end does not give a real right to anything further than barely the means necessary to the attainment of that end.

---

* Vattel, B. III, chap. 8, sec. 145.

† Vattel, B. III, chap. 5, sec. 72.

‡ Vattel, B. III, chap. 8, sec. 147. Incorporated by Halleck, Law of War, chap. 18, sec. 3.

§ Vattel, B. III, chap 8, sec. 138.

Whatever we do beyond that is reprobated by the law of nature, is faulty and condemnable at the tribunal of conscience. Hence it is that the right to such, or such acts of hostility, varies according to circumstances. What is just and perfectly innocent in war in one particular situation, is not always so on other occasions. Right goes hand-in-hand with necessity and the exigencies of the case, but never exceeds them." *

"All these classes (old men, women and children, the clergy, magistrates, and other civil officers), which, by general usage or the municipal laws of the belligerent State, are exempt from military duty, are not subject to the general rights of a belligerent over the enemy's person. To these are added, by modern usage, all persons who are not organized or called into military service, though capable of its duties, but who are left to pursue their usual pacific avocations. All these are regarded as *non-combatants*."†

General Sherman admits, in his Memoirs, that he burned stores and dwellings; that "the heart of the city was in flames all night;" that he telegraphed to Grant he had "made a wreck of Atlanta,"‡ which he afterwards termed "the ruined city." The following quotations will show whether or not he was justified in this destruction of property:

"And with respect to things, the case is the same as with respect to persons—things belonging to the enemy, continue such wherever they are. But we are not hence to conclude, any more than in the case of persons, that we everywhere possess a right to treat these things as things belonging to the enemy." §

"The wanton destruction of public monuments, temples, tombs, statues, paintings, etc., is absolutely condemned, even by the voluntary law of nations, as never being conducive to the lawful object of war. The pillage and destruction of towns, the devastation of the open country, ravaging, setting fire to houses, are measures no less odious and detestable on every occasion where they are evidently put in practice without absolute necessity or, at least, very cogent reasons. But as the perpetrators of such outrageous deeds might attempt to palliate them, under pretext of deservedly punishing the enemy, be it here observed that the natural and voluntary law of nations does not allow us to inflict

---

* Vattel, B. III, chap. 8, sec. 137,

† Halleck, Laws of War, chap. 16, sec. 2.

‡ Sherman's Memoirs, vol. II, page 154.

§ Vattel, B. III, chap. 5, sec. 74.

such punishments, except for enormous offences against the law of nations."*

When General Lee entered Pennsylvania with his Army, he gave strict orders to destroy no property, and to pay for all provisions obtained from the enemy. Marshal Soult was likewise magnanimous in his conduct, after he had been not only compelled to storm the defences of Oporto, but to fight from street to street, in order to finally force a surrender. Napier states that the French found some of their comrades who had been taken prisoners, "fastened upright and living, but with their eyes burst, their tongues torn out, and their other members mutilated and gashed." This ghastly sight notwithstanding, many of the French soldiers and officers endeavored, at the risk of their lives, to check the vengeance of their comrades, Soult did not, even after this fearful resistance and these examples of barbarous cruelty, send off the women and children, the infirm and the sick, and then burn their homes; on the contrary,† "Recovering and restoring a part of the plunder, he caused the inhabitants remaining in town to be treated with respect; he invited, by proclamation, all those who had fled to return, and he demanded no contribution; but restraining with a firm hand the violence of his men, he contrived, from the captured public property, to support the Army and even to succor the poorest and most distressed of the population."

Although it is customary, previous to a general assault of a fortified town of which the demand for surrender has been rejected, that the commanding officer give warning (on account of the extraordinary sacrifice of life, to which his troops must necessarily be subjected) that he will not be responsible for the lives of the captured, as did Lieutenant General Lee in my name at Resaca. No officer should allow his soldiers to burn and pillage after victory has been secured.

---

* Vattel, B. III, chap. 9, sec. 173. Incorporated by Halleck. Laws of War, chap. 19, sec. 24.

† Napier, Peninsular War, B. VI, chaps. 4 and 7.

CHAPTER XV.

CAMPAIGN TO THE ALABAMA LINE—PREPARATIONS—PRESIDENT DAVIS VISITS THE ARMY—HARDEE RELIEVED—ALLATOONA —RESACA—DALTON—LAFAYETTE—SUDDEN DETERMINATION TO ENTER TENNESSEE—GADSDEN—BEAUREGARD.

AFTER the fall of Atlanta, this most serious question presented itself for solution: in what manner, and accompanied with the least detriment, to effect the riddance of a victorious foe, who had gained possession of the mountains in our front, and planted his standard in the heart of the Confederacy. In order to compass this end, either the Federals should be forced back by manœuvres into the mountains, there defeated in battle, and finally driven northward; or an attempt be made to defeat them upon their march forward, *after* Sherman had been allowed full time to rest his troops, make preparations, and receive reinforcements, for, in the meantime, it would have been rashness and folly, in view of our inferior numbers, to have attacked the enemy whilst under the protection of the breastworks of Atlanta. This grave and momentous question presented the same difficulties which had risen before General Lee, when Grant crossed the Rappahannock, and the battles of the Wilderness, Spottsylvania Court House, and, finally, the surrender at Appomattox followed. Our great chieftain well knew that he would be forced to abandon Richmond or surrender his Army, unless he beat his enemy in battle, and drove him back, as he had done in previous instances. So paramount did he consider this necessity that he cut roads through the Wilderness, in order to get at the Federals while his own Army was in best condition for battle. He possessed, for the execution of his purpose, a body of troops which had been trained and handled in such a manner as to render it impossible to find its superior in the history of nations. Moreover, he was not confronted by a victorious Army, but by one he had driven back more than once from the same line then occupied by Grant.

The difficulties which surrounded me even at the outset, when I assumed command of the Army of Tennessee, and after the fall of Atlanta, when a *recurrence* of retreat was brought about with its train of former evils, were more perplexing than those which beset General Lee at the juncture above referred to. The problem was the more difficult to solve, by reason of the impaired condition of the Army.

The same question had arisen for consideration when Sherman moved from Chattanooga, and formed line of battle in front of Rocky-faced Ridge. My predecessor did not perceive the necessity of defeating the enemy at that period—a necessity as urgent as that which impelled General Lee to use extraordinary means to reach his enemy in the Wilderness.

Unless the Army could be heavily reinforced, there was, in the present emergency, but one plan to be adopted: by manœuvres to draw Sherman back into the mountains, then beat him in battle, and at least regain our lost territory. Therefore, after anxious reflection, and consultation with the corps commanders, I determined to communicate with the President, and ascertain whether or not reinforcements could be obtained from any quarter. In accordance with this decision, I telegraphed to General Bragg as follows:

[No. 1.]

"LOVEJOY STATION, *September 3d, 1.45 p. m.*

"For the offensive my troops, at present, are not more than equal to their own numbers. To prevent this country from being overrun, reinforcements are absolutely necessary.

"J. B. HOOD, *General.*"

At 6.10 p. m., the same day:

[No. 2.]

"My telegram in cipher this morning is based upon the supposition that the enemy will not content himself with Atlanta, but will continue offensive movements. All the Lieutenant Generals agree with me.

"J. B. HOOD, *General.*"

In consideration of the high regard President Davis entertained for General Hardee, I suggested to the latter to telegraph to the President in relation to our condition. I find in my dispatch book a copy of his telegram:

[No. 3.]

"*September 4th, 11.30 a. m.*

"Unless this Army is speedily and heavily reinforced, Georgia and Alabama will be overrun. I see no other means to arrest this calamity. Never, in my opinion, were our liberties in such danger. What can you do for us?

"(Signed) W. J. HARDEE, *Lieutenant General.*"

The following reply from His Excellency conveyed no hope of assistance:

"RICHMOND, *September 5th, 1864.*

"GENERAL J. B. HOOD:—Your dispatches of yesterday received. The necessity for reinforcements was realized, and every effort made to bring forward reserves, militia, and detailed men for the purpose. Polk, Maury, S. D. Lee, and Jones have been drawn on to fullest extent. E. K. Smith has been called on. No other resource remains. It is now requisite that absentees be brought back, the addition required from the surrounding country be promptly made available, and that the means in hand be used with energy proportionate to the country's need.

"JEFFERSON DAVIS."

I hereupon decided to operate at the earliest moment possible in the rear of Sherman, as I became more and more convinced of our inability to successfully resist an advance of the

Federal Army. I had thought immediately after my arrival at Lovejoy Station that our troops were not disheartened, and telegraphed to Richmond to that effect; but I discovered my error before long, and concluded to resume active operations, move upon Sherman's communications, and avert, if possible, impending disaster from the Confederacy.

Before entering into the details of the plan of the contemplated campaign, I will, in brief, consider the indubitable results had I remained in front of Sherman, till he made ready and moved forward. In lieu of dividing his forces, as he did when I eventually marched to his rear, he would either have increased the strength of his Army to the fullest extent possible, previous to his forward movement, in order not only to brush away more easily the cobweb of an Army in his front, but also to overawe and discourage our people by the presence of an Army strong and powerful; or he would have ordered Thomas into Tennessee, with instructions to muster all available forces and march into Alabama with a second Army, whilst he moved through Georgia. In the event of the adoption of the first plan, he could, after assembling all the troops at his disposal between Nashville and Atlanta, have advanced with an Army of not less than one hundred and twenty-five thousand (125,000). According to his own statement, Thomas had under his command, at the time I accepted battle at Nashville, over seventy thousand (70,000) effectives, irrespective of troops at other points in Tennessee and Kentucky.

Had he chosen the second plan, he would soon have moved with a concentrated Army of not less than seventy-five thousand (75,000), whilst Thomas overran Alabama with at least fifty thousand (50,000) men. This is in no degree an exaggerated estimate, since forces could have been withdrawn from Tennessee and Kentucky, where no necessity for troops would have existed during these operations. The enthusiasm throughout the North, succeeding the capture of Atlanta, would also have swollen the Federal ranks by the return of absentees in large numbers. This plan would have brought into the field two powerful Armies to move simultaneously through Georgia and Alabama.

On the other hand, our Army of forty thousand four hundred and three (40,403) would have gradually decreased through desertions, with no prospect of obtaining another man east of the Mississippi river, and with the information in my possession from Richmond, that no troops were shortly expected from the Trans-Mississippi Department, although every effort had been made by the Government to get reinforcements from that quarter.

Thus the outcome of this stand-still policy, which would have enabled Sherman to advance with all due preparations and have forced us to retreat in his front day after day, would have been the final dispersion of the Army; a greater portion would have returned to their homes, leaving behind a noble band of patriots too proud to desert, yet too weak and disheartened to be of material service. I would have been able to offer just about sufficient resistance to harass and embitter the enemy; to instigate him to perpetrate greater outrages, and commit ten-fold the havoc he actually made in traversing Georgia; and, in lieu of contenting himself with simply cutting the communications of the Army of Northern Virginia, with its largest fields for supplies, Sherman would have tarried long enough upon his march to effect irreparable damage.

I shall now recite the preliminaries to the campaign in rear of Sherman, and give an account of operations, results accomplished, together with those events which led me to conceive the idea of the campaign into Tennessee.

I foreshadowed my intention of moving upon Sherman's communications in the following telegram to the President, dated September 6th:

[No. 10.]

"I shall make dispositions to prevent the enemy, as far as possible, from foraging south of Atlanta, and at the same time endeavor to prevent his massing supplies at that place. I deem it important that the prisoners at Andersonville should be so disposed of, as not to prevent this Army from moving in any direction it may be thought best. According to all human calculations, we should have saved Atlanta had the officers and men of the Army done what was expected of them. It has been God's will for it to be otherwise. I am of good heart, and feel that we shall yet succeed. The Army is much in need of a little rest. After removing the prisoners from Andersonville, I think we should, as soon as practicable, place our Army upon the communications of the enemy, drawing our supplies from the West Point and Montgomery Railroad. Looking to this, I shall at once proceed to strongly fortify Macon. Please do not fail to give me advice at all times. It is my desire to do the best for you and my country. May God be with you and us.

"J. B. HOOD, *General.*"

Having requested and obtained authority from the War Department to propose an exchange of prisoners, captured during the siege, I made on the 8th of September, by flag of truce, a proposition to the enemy to that effect. An exchange of two thousand (2000) was agreed upon. Some delay, however, resulted from a refusal upon the part of General Sherman to excnange Confederates for Federal prisoners whose term of service had ceased or was about to expire.

Upon the 9th was initiated the correspondence between General Sherman and myself, in regard to the treatment of the inhabitants of Atlanta, and which I embodied in the narrative of the siege of that city. On the 12th I sent every wagon, which could be spared in the Army, to Rough and Ready, and performed the sad duty of transferring within our lines the women and children, the sick and the infirm.

In the meantime, intelligence had been received from General Wheeler, announcing that he had destroyed several bridges and about fifty miles of railroad in Tennessee, and that he had thus far been successful in every engagement with the enemy.

During the progress of the exchange of prisoners, the transportation of the Army was carefully inspected and repaired; pontoon trains made ready for active operation, and every exertion made to inaugurate a forward movement at the earliest hour possible. At this period I deemed it to the interest of the Confederacy, because of General Hardee's failure to obey instructions on the 20th, and 22d of July, and 31st of August, to request that this officer be relieved from duty with his Corps, and that another be assigned to its command. I dispatched to General Bragg as follows:

[No. 14.]

"*September 8th, 2.30 p. m.*

"I suggest that all the reserves of Georgia, under General Cobb, be ordered to this Army, since the prisoners have been removed; and that Lieutenant General Taylor be ordered to relieve General Hardee, bringing with him all the troops he can.

"J. B. HOOD, *General.*"

The unfortunate events connected with General Hardee's service during the siege of Atlanta, rendered obligatory this unpleasant duty on my part. I have already stated the opinions, at the time, of Lieutenant Generals Lee and Stewart, and of Major General G. W. Smith, in regard to this painful subject. So decided were these officers in their convictions that I determined to inform the President of my own loss of confidence, and to invite him to visit the Army, and confer with the corps commanders in relation to the operations around Atlanta. Accordingly, I sent the following telegram to His Excellency:

[No. 24.]

"*September 13th.*

"In the battle of July 20th, we failed on account of General Hardee. Our success on the 22d July was not what it should have been, owing to this officer. Our failure on the 31st of August, I am convinced, was greatly owing to him. Please confer with Lieutenant Generals Stewart and S. D. Lee, as to operations around Atlanta. It is of the utmost importance that Hardee should be relieved at once. He commands the best troops of this Army. I must have another commander. Taylor or Cheatham will answer. Hardee handed in his resignation a few days since, but withdrew it. Can General Cobb give me all the reserve regiments he has?

"J. B. HOOD, *General.*"

Major General Gustavus W. Smith, in his official report of the operations of the Georgia State troops, dated 15th of

September, 1864, shortly after these occurrences, says in this regard:

"Commanding a peculiar organization, the ranking officer of the forces of the State in which you were operating, I was invited to and participated in your councils. I had every opportunity of knowing what was going on. Your plans were fully explained to your Lieutenant Generals, your chief of artillery, chief engineer, and myself. Opinions and views were called for, and then specific orders were given. I have never known one of them to dissent to any plan of yours, a doubt expressed as to the meaning, or intent, of your orders, nor a suggestion made by them of a plan they supposed would be better than that you ordered. If they are not now unanimous, there is but *one*, if any, who dissents from the opinion expressed above, viz: Sherman would have been beaten, had your orders been obeyed on the 20th July, 22d July, and 31st August."

General A. P. Stewart, in his official report of the operations around Atlanta, states in regard to the battle of the 20th July:*

"I cannot but think had the plan of the battle, as I understood it, been carried out fully, we would have achieved a great success."

Lieutenant General Lee expressed to me the opinion that but for the delay before the attack on the 31st August, the result might have been different. This officer, Lieutenant General A. P. Stewart, and Major General G. W. Smith were, at the time, unanimous in the conviction that had General Hardee faithfully and earnestly carried out my instructions on the 20th, and 22d July, we would have been victorious in the two battles, *i. e.*, had he attacked at 1 o'clock in lieu of 4 p. m., on the 20th; had he appealed to his troops in a manner to arouse their pride, patriotism and valor, instead of giving utterance to expressions of caution against breastworks; had he, on the 22d, marched *entirely* round and in rear of McPherson's left flank, as ordered, and attacked at daylight or early morning, we would have gained signal victories.

It may very properly be asked why, after failure on two consecutive occasions, was Hardee placed in command at

---

* Appendix, page 350.

Jonesboro'; why I did not relieve him previously from duty with the Army, and thus avoid further cause of complaint.

The battles of the 20th, and 22d of July, were fought in rapid succession, and immediately after my appointment to the command of the Army. I knew not then the original cause of trouble, nor was I enlightened upon this matter till General Cleburne visited my headquarters about two weeks after these engagements. The President had confidence in General Hardee, and believed he could be of great service on account of his thorough knowledge of the country, and his long connection with the Army of Tennessee. In this opinion I naturally acquiesced, since I could not imagine that a soldier, wittingly and willingly, would disregard orders in operations of so much importance. Moreover, the position of his line of battle, together with that of General Lee, rendered it necessary to send their two corps to Jonesboro', and Hardee, the superior officer in rank, of course assumed command.

I was slow and reluctant to adopt the conclusion finally expressed in my dispatch to the President. I refused to attribute Hardee's non-fulfilment of orders to a fixed purpose on his part to thwart my operations as Commander-in-Chief, and imputed his misfortune mainly to the influence of the school in contact with which he had been thrown for a considerable period.

It is true I had been promoted and placed over him who was my senior in rank, and equally true that, under similar circumstances, not many men will co-operate as heartily as duty dictates, in the furtherance of the projects of their commanding officer. His brother corps commanders were of the opinion that in this grievance lay the source of trouble, and that he, if not consenting to a frustration of my plans, was at least willing I should not achieve signal success. If these impressions be correct, his want of confidence, or rather fear of rashness on my part, was not lessened; and feelings were doubtless engendered that created the lukewarmness which characterized the conduct of his military operations at that juncture.

I have been forced to recur to these facts, in consequence of their intimate connection with the important events of that period, and do so with the more sincere regret that General Hardee is no longer able to speak in his own defence.

After the removal of the prisoners at Andersonville—hitherto the principal obstacle to a movement in rear of Sherman—I deemed it advisable, and, therefore ordered that the railroad iron for some distance on the three roads leading into Atlanta, be removed and stored for future use. Major General M. L. Smith, chief engineer, was instructed to not only fortify Macon, but likewise Augusta and Columbus; the chief commissary was directed to remove the depot of supplies to the West Point Railroad, as I desired, preparatory to crossing the Chattahoochee, to place our left flank on that river, with headquarters at Palmetto.

I recalled General Wheeler from Tennessee to join immediately the left of the Army, whilst Colonel Prestman, of the engineer corps, made ready to move with the pontoon train and a sufficient number of boats to meet any emergency. These various preparations somewhat revived the spirit of the officers and men; I was hereby induced to believe that the Army, in its next effort at battle, would fight with more determination than had been exhibited since our retreat from Resaca, and so telegraphed General Bragg on the 15th of September.

Upon the morning of the 18th, the Army began to move in the direction of the West Point Railroad, which the advance reached on the 19th. Upon the 20th, line of battle was formed, with the right east of the railroad, and the left resting near the river, with Army headquarters at Palmetto.

I sent the following dispatch to General Bragg the succeeding day:

[No. 30.]

"*September 21st.*

"I shall—unless Sherman moves south—as soon as I can collect supplies, cross the Chattahoochee river, and form line of battle near Powder

Springs. This will prevent him from using the Dalton Railroad, and force him to drive me off or move south, when I shall follow upon his rear. I make this move as Sherman is weaker now than he will be in future, and I as strong as I can expect to be. Would it not be well to move a part of the important machinery at Macon to the east of the Oconee river, and do the same at Augusta to the east side of the Savannah river? If done, it will be important to make the transfer so as not to interfere with the supplies for the Armies.

"J. B. HOOD, *General.*"

On this date expired the truce of ten days which had been agreed upon for the exchange of prisoners, and Major Clare, of my staff, returned with his escort from Rough and Ready. The same day I received information that the President, in response to my invitation, had decided to visit the Army forthwith.

On the 25th, at 3.30 p. m., President Davis, accompanied by two staff officers, arrived at Palmetto, with a view to ascertain in person the condition of the Army; to confer, as requested, with the corps commanders in regard to the operations around Atlanta, and to obtain the particulars of the proposed campaign in the rear of Sherman.

On the ensuing morning, we rode forth together to the front, with the object of making an informal review of the troops. Some brigades received the President with enthusiasm; others were seemingly dissatisfied, and inclined to cry out, "give us General Johnston." I regretted I should have been the cause of this uncourteous reception to His Excellency; at the same time, I could recall no offence save that of having insisted that they should fight for and hold Atlanta forty-six days, whereas they had previously retreated one hundred miles within sixty-six days.

During the evening the President was serenaded by the Twentieth Louisiana band, accompanied by quite a large number of soldiers. He made upon the occasion a short but spirited speech, which was received with long and continued cheers. General Howell Cobb, and Governor Harris, of Tennessee, also delivered brief and eloquent addresses.

The President held a long conference the next day with Lieutenant Generals Lee and Stewart, in a house not far from my tent. He conferred also separately with General Hardee. I had, at a previous interview, fully expressed to him my views in relation to the condition of the Army, and maintained that our only hope to checkmate Sherman was to assume the offensive, cut the enemy's communications, select a position on or near the Alabama line in proximity to the Blue Mountain Railroad, and there give him battle. Should the enemy move south, I could as easily from that point as from Palmetto, follow upon his rear, if that policy should be deemed preferable. On the other hand, if my position on or near the Alabama line should force Sherman to move out of Atlanta—as I believed it would do—and divide his Army by sending off a portion to Tennessee, which he would consider immediately threatened, I might be able to defeat the wing of the Federal Army, remaining in Georgia, drive it from the country, regain our lost territory, reinspirit the troops, and bring hope again to the hearts of our people. I stated also that I thought an offensive move would improve the *morale* of the Army to a degree which would render it equal to giving battle to the enemy; that at the moment it was totally unfit for pitched battle, and the above plan offered the sole chance to avert disaster.

Prior to his departure, I recalled to him the fact that I had accepted with reluctance the position to which I had been assigned; that I had never sought preferment from him either directly or indirectly, and assured him I cherished but one desire, which was to do my whole duty to my country. I told him I was aware of the outcry against me, through the press, since the removal of Johnston, and, if he adjudged a change of commanders expedient, not to hesitate to relieve me entirely from duty with the Army of Tennessee or to give me a corps or division, under a more competent leader than myself.

After final counsel with the Lieutenant Generals, he left for Montgomery, at 6 p. m., on the 27th of September.

The main part of the above conversation was repeated after he had mounted his horse, and was in readiness to leave; he replied that he might find it necessary to assign another to the command of the Army, but I should continue to pursue my proposed plan, at least till a decision was reached in the matter.

On the 28th, an order from the President was received, and read to the troops, relieving Lieutenant General Hardee from duty with the Army of Tennessee, and assigning him to the command of the Department of South Carolina and Florida.

The same day, I issued instructions to commence the movement across the Chattahoochee at Pumpkin Town and Phillips's Ferry, and, on the following morning, I directed that our supplies from Newnan cross the river at Moore's Ferry. At noon, I rode over the pontoon bridge in advance of the infantry, and established my headquarters that night at Pray's Church, along with General Jackson, commanding the cavalry; and on the next day I received the subjoined communication from the President:

[Private.]

"OPELIKA, ALABAMA, *September 28th, 1864.*

"GENERAL JOHN B. HOOD, *Headquarters Army of Tennessee.*

"GENERAL:—I have anxiously reflected upon the subject of our closing conversation and the proposition confidentially mentioned. It seems to me best that I should confer with General Beauregard, and, if quite acceptable to him, place him in command of the department embracing your Army and that of General R. Taylor, so as to secure the fullest co-operation of the troops, without relieving either of you of the responsibilities and powers of your special commands, except in so far as would be due to the superior rank and the above assignment of General Beauregard. He would necessarily, if present with either Army, command in person. Before final action, there will be time for you to communicate with me, and I shall be glad to have your views. In the meantime you will of course proceed as though no modification of existing organization was contemplated.

"Very respectfully and truly yours,

"JEFFERSON DAVIS."

The morning of the 1st of October, Brigadier General Jackson advanced with the cavalry, sending a detachment at the same time to operate against the railroad between the Chattahoochee and Marietta. That night the Army went into bivouac eight miles north of Pray's Church, after having effected an undisturbed and safe passage of the Chattahoochee. Information was here received that Kilpatrick's cavalry was north of the river, and that Girard's cavalry had moved in the direction of Rome.

The next morning, I telegraphed to General Bragg as follows:

[No. 33.]

"*October 2d.*

"To-night my right will be at Powder Springs, with my left on Lost Mountain. This will, I think, force Sherman to move on us or to move south. Should he move towards Augusta, all available troops should be sent *there* with an able officer of high rank to command. Could General Lee spare a division for that place in such an event?

"J. B. HOOD, *General.*"

The night of the 2d, the Army rested near Flint Hill Church. On the morning of the 3d, Lieutenant General Stewart was instructed to move with his Corps, and take possession of Big Shanty; to send, if practicable, a detachment for the same purpose to Ackworth, and to destroy as great a portion of the railroad in the vicinity as possible; also to send a division to Allatoona to capture that place, if, in the judgment of the commanding officer, the achievement was feasible.* The main body of the Army in the meantime moved forward, and bivouacked near Carley's house, within four miles of Lost Mountain.

On the 4th, General Stewart captured, after a slight resistance, about one hundred and seventy prisoners, at Big Shanty, and, at 9.30 a. m., the garrison at Ackworth, numbering two hundred and fifty men, surrendered to General Loring. The forces under these officers joined the main body near Lost

---

* See Official Report, Appendix, page 326.

Mountain on the morning of the 5th, having, in addition, destroyed about ten or fifteen miles of the railroad.

I had received information—and General Shoupe records the same in his diary—that the enemy had in store, at Allatoona, large supplies which were guarded by two or three regiments. As one of the main objects of the campaign was to deprive the enemy of provisions, Major General French was ordered to move with his Division, capture the garrison, if practicable, and gain possession of the supplies. Accordingly, on the 5th, at 10 a. m., after a refusal to surrender, he attacked the Federal forces at Allatoona, and succeeded in capturing a portion of the works; at that juncture, he received intelligence that large reinforcements were advancing in support of the enemy, and, fearing he would be cut off from the main body of the Army, he retired and abandoned the attempt. Major L. Perot, adjutant of Ector's brigade, has informed me by letter that our troops were in possession of these stores during several hours, and could easily have destroyed them. If this assertion be correct, I presume Major General French forbade their destruction, in the conviction of his ability to successfully remove them for the use of the Confederate Army.

Our soldiers fought with great courage; during the engagement Brigadier General Young, a brave and efficient officer, was wounded, and captured by the enemy.

General Corse won my admiration by his gallant resistance, and not without reason the Federal commander complimented this officer, through a general order, for his handsome conduct in the defence of Allatoona.

Our presence upon his communications compelled Sherman to leave Atlanta in haste, and cross the Chattahoochee on the 3d and 4th of October with, according to our estimate at that time, about sixty-five thousand (65,000) infantry and artillery, and two divisions of cavalry. He left one corps to guard the city and the railroad bridge across the river, and telegraphed to Grant he would attack me if I struck his road south of the Etowah.

I received at this juncture a copy of the following order from President Davis:

"AUGUSTA, GEORGIA, *October 2d, 1864.*

"GENERAL G. T. BEAUREGARD, *Augusta, Georgia.*

"GENERAL:—I desire that, with as little delay as practicable, you will assume command of the military departments now commanded respectively by General Hood, and Lieutenant General Taylor.

"You will establish the headquarters of the department under your command at such point within its territorial limits as you may consider most advantageous to the public service.

"Your personal presence is expected wherever in your judgment the interests of your command render it expedient, and whenever present with an Army in the field, you will exercise immediate command of the troops.

"The adjutant and inspector general will be directed to communicate to you without delay the orders defining the geographical limits of your department, and such letters of general instruction as may have been sent to your predecessors, and as it may be important for you to possess.

"Very respectfully and truly yours,

"JEFFERSON DAVIS.

"(For GENERAL HOOD)."

This order was most satisfactory, inasmuch as it afforded me at least an opportunity to confer with an officer of distinction, in regard to future operations.

The attack upon his communications, in the vicinity of the Etowah river and near the Alabama line, had forced Sherman to hasten from Atlanta. In truth, the effect of our operations so far surpassed my expectations that I was induced to somewhat change my original plan to draw Sherman to the Alabama line and then give battle. I accordingly decided to move further north and again strike his railroad between Resaca and Tunnel Hill, thoroughly destroy it, and then move in the direction of the Tennessee, via Lafayette and Gadsden, with no intent, however, to cross the river. This move, I considered, would so seriously threaten the road at Stevenson, and the bridge across the Tennessee river, at Bridgeport, that Sherman would be compelled still further to detach and divide his forces, whilst at the same time he continued his march northward. I intended then to entice him as near the Tennessee line as possible, before offering battle. To accomplish this end, I thought it might be expedient to march to a point in the vicinity of the Tennessee, and even to order the cavalry to advance as far as the river, before I turned upon the enemy.

It was my fixed purpose to attack Sherman as soon as I succeeded in these manœuvres. The plan of the original campaign was, therefore, only more fully developed by this strategy, which, in truth, I adopted as an afterthought.

On the 6th, the Army reached Dallas; our right rested at New Hope Church, where intelligence was received that the enemy was advancing from Lost Mountain. From Dallas we marched to Coosaville, ten miles southwest of Rome, via Van Wert, Cedartown, and Cave Spring. At the latter place Major General Wheeler, with a portion of his command, joined me from Tennessee. We arrived at Coosaville on the 10th, and the day previous, when near Van Wert, I sent the following dispatch to General Bragg:

[No. 34.]

"NEAR VAN WERT, GEORGIA,
"*October 9th, 1864.*

"GENERAL B. BRAGG AND HONORABLE J. A. SEDDON, *Richmond.*

"When Sherman found this Army on his communications, he left Atlanta hurriedly with his main body, and formed line of battle near Kennesaw Mountain. I at once moved to this point, and, marching to-morrow, shall cross the Coosa river about ten miles below Rome; and moving up the west bank of the Oostenaula, hope to destroy his communications from Kingston to Tunnel Hill, forcing him to fall back or move south. If the latter, I shall move on his rear. If the former, I shall move to the Tennessee river, via Lafayette and Gadsden. I leave near Jacksonville all surplus baggage, artillery, and wagons, and move prepared for battle. If I move to the Tennessee, my trains will meet me at Gadsden. Please have the Memphis and Charleston Railroad repaired at once to Decatur, if possible.

"J. B. HOOD, *General.*"

This last precautionary measure I deemed advisable, as I sought to forestall every possible contingency. If our arms met with only partial success in battle—that is, if Sherman was not routed, but merely badly worsted—I had determined to send the wounded to the rear by the Blue Mountain Railroad; by rapid marches to cross the Tennessee river at Gunter's Landing, and again destroy the enemy's communications at Stevenson, and Bridgeport. I felt confident that Sherman, after being disabled in battle, would follow in my rear, and I hoped that the near approach of cold weather would favor my attempt to at least recover our lost territory, and allow our Army to winter again in the vicinity of Dalton. In anticipation of this probable event, I requested the authorities to have the Memphis and Charleston Railroad repaired to or near Decatur, Alabama, in order to establish another line for supplies and retreat, in case of either success or disaster in Tennessee.

In a dispatch to General Taylor I requested that Forrest be ordered to operate at once in Tennessee:

[No. 499.]

"VAN WERT, *October 7th.*

"LIEUTENANT GENERAL TAYLOR,

"*Commanding Department, Gainesville Junction.*

"Your dispatch of the 6th received. This Army being in motion, it is of vital importance that Forrest should move without delay, and operate on the enemy's railroad. If he cannot break the Chattanooga and Nashville Railroad, he can occupy their forces there, and prevent damage being repaired on the other road. He should lose no time in moving. I am very thankful for the assistance already afforded this Army.

"J. B. HOOD, *General.*"

The improvement in the *morale* of the troops was already apparent, and desertions, so frequent at Palmetto, had altogether ceased. I, therefore, indulged a not unreasonable hope very soon to deal the enemy a hard and staggering blow. In order to convey his appreciation of the importance attached to our movement upon his line of communication, I will quote General Sherman's own words:*

"In person I reached Allatoona on the 9th of October, still in doubt as to Hood's immediate intentions."

---

* Sherman's Memoirs, vol. II, page 152.

In a dispatch of the same date to Thomas, at Nashville:

"I came up here to relieve our road. The Twentieth Corps remains at Atlanta. Hood reached the road, and broke it up between Big Shanty and Ackworth. He attacked Allatoona, but was repulsed. We have plenty of bread and meat, but forage is scarce. I want to destroy all the road below Chattanooga, including Atlanta, and to make for the sea coast. We cannot defend this long line of road."

On the same day he sent the following dispatch to Grant, at City Point:

"It will be a physical impossibility to protect the roads now that Hood, Forrest, Wheeler, and the whole batch of devils are turned loose without home or habitation. I think Hood's movements indicate a diversion to the end of the Selma and Talladega road, at Blue Mountain, about sixty miles southwest of Rome, from which he will threaten Kingston, Bridgeport, Decatur, Alabama." * * * *

On the 10th of October, Brigadier General Jackson, commanding the cavalry, was instructed by Colonel Mason, as follows:

[No 438.]

"CAVE SPRING, *October 10th, 8 a. m.*

"General Hood desires me to inform you that the pontoon at Quinn's Ferry, on the Coosa river, will be taken up this evening, and you must put on a line of couriers to that place to connect with a line to the other side. They will meet at the ferry, and you must continue to keep some there, or near there, to take dispatches over the line. Day after to-morrow (12th), unless you are otherwise engaged, General Hood desires you will move on Rome, and make a considerable demonstration from your side of the river; but be careful not to fire into the town. Communicate fully and frequently about all movements of the enemy."

On the 11th, the Army crossed the Coosa river, marched in the direction of Resaca and Dalton, and bivouacked that night fourteen miles above Coosaville, and ten miles northwest of Rome. That same day Major General Arnold Elzey, chief of artillery, was directed to move to Jacksonville with the reserve artillery and all surplus wagons, and General Jackson was instructed to retard the enemy as much as possible, in the event of his advance from Rome.

Having thus relieved the Army of all incumbrance, and

made ready for battle, we marched rapidly to Resaca, and thence to Dalton, via Sugar Valley Post Office. Lieutenant General Lee moved upon Resaca, with instructions to display his forces and demand the surrender of the garrison, but not to attack, unless, in his judgment, the capture could be effected with small loss of life. He decided not to assault the Federal works, and commenced at once the destruction of the railroad.

On the 13th, I demanded the surrender of Dalton, which, in the first instance, was refused, but was finally acceded to at 4 p. m. The garrison consisted of about one thousand (1000) men. As the road between Resaca and Tunnel Hill had been effectually destroyed, the Army was put in motion the next morning in the direction of Gadsden, and camped that night near Villanon.

Whilst in front of Dalton, quite a spirited affair occurred at Mill Creek Gap, where a detachment of our troops attacked and gained possession of a block house. Major Kinloch Falconer, of my staff, was during this assault seriously wounded.

On the morning of the 15th, I sent from Van Wert the following dispatch to the Honorable J. A. Seddon, Secretary of War, Generals Bragg and Beauregard:

[No. 500.]

"VAN WERT, *October 15th.*

"This Army struck the communications of the enemy about a mile above Resaca (the 12th), completely destroying the railroad, including block houses from that point to within a short distance of Tunnel Hill; and about four miles of the Cleveland Railroad, capturing Dalton and all intermediate garrisons, with their stores, arms and equipments—taking about one thousand (1000) prisoners. The main body of Sherman's Army seem to be moving towards Dalton.

"J. B. HOOD, *General.*"

From Villanon, the Army passed through the gaps in the mountains, and halted during the 15th and 16th at Cross Roads, in a beautiful valley about nine miles south of Lafayette. At this time I received intelligence that Sherman had, on the 13th, reached Snake Creek Gap, where the right of his line had rested in the early Spring of this year; also that he was marching in our pursuit, whilst General Wheeler was endeavoring to retard his advance as much as possible. I here determined to advance no further towards the Tennessee river, but to select a position and deliver battle, since Sherman had, at an earlier date than anticipated, moved as far north as I had hoped to allure him; moreover I was again in the vicinity of the Alabama line, with the Blue Mountain Railroad in my rear, and I thought I had discovered that improvement in the *morale* of the troops, which would justify me in delivering battle. In accordance with information received from our cavalry, Sherman had, however, made no further division of his forces after leaving Atlanta. I estimated, therefore, his strength to be about sixty-five thousand (65,000) effectives.

Upon the eve of action, I considered it important to ascertain by personal inquiry and through the aid of officers of my staff, not alone from corps commanders, but from officers of less rank, whether or not my impressions after the capture of Dalton were correct, and I could rely upon the troops entering into battle at least hopeful of victory. I took measures to obtain likewise the views of Lieutenant General Lee who, at this juncture, was with his Corps in rear, at or near Ship's Gap. He agreed with all the officers consulted; the opinion was unanimous that although the Army had much improved in spirit, it was not in condition to risk battle against the numbers reported by General Wheeler.

The renouncement of the object for which I had so earnestly striven, brought with it genuine disappointment; I had expected that a forward movement of one hundred miles would re-inspirit the officers and men in a degree to impart to them confidence, enthusiasm, and hope of victory, if not strong faith in its achievement.

I remained two days at Cross Roads in serious thought and perplexity. I could not offer battle while the officers were *unanimous* in their opposition. Neither could I take an entrenched position with likelihood of advantageous results, since Sherman could do the same, repair the railroad, amass a large Army, place Thomas in my front in command of the forces he afterwards assembled at Nashville, and then, himself, move southward; or, as previously suggested, he could send Thomas into Alabama, whilst he marched through Georgia, and left me to follow in his rear. This last movement upon our part, would be construed by the troops into a retreat, and could but result in disaster.

In this dilemma, I conceived the plan of marching into Tennessee with a hope to establish our line eventually in Kentucky, and determined to make the campaign which followed, unless withheld by General Beauregard or the authorities at Richmond. General Beauregard at this time was journeying in my direction. I proposed, therefore, when he joined me, to lay fully before him my plan of operations.

Before entering into an account thereof, I will for a moment advert to the evidences of the solicitude occasioned the enemy by our movement to the Alabama line.

On the 10th of October, General Sherman telegraphed to Generals Thomas and Cox, as follows: *

"I will be at Kingston to-morrow. I think Rome is strong enough to resist any attack, and the rivers are all high. If he (Hood) turns up by Summerville, I will get behind him."

On the 16th, when in pursuit of our Army from Resaca in the direction of Ship's Gap and Lafayette, he again telegraphs to Thomas, at Nashville: †

"Send me Morgan's and Newton's old Divisions. Re-establish the road, and I will follow Hood wherever he may go. I think he will move to Blue Mountain. We can maintain our men and animals on the country."

On the 17th, he writes Schofield, at Chattanooga: ‡

"* * * We must follow Hood till he is beyond the reach of mischief, and then resume the offensive."

---

* Sherman's Memoirs, vol. II, page 153.

† Sherman's Memoirs, vol. II, page 156.

‡ Sherman's Memoirs, vol. II, page 157.

Ten days after this declaration, he was still undecided as to the plan he should adopt. In truth, it seemed difficult to divine when our little Army would be far enough away to be "beyond the reach of mischief." On the 26th, he telegraphed to General Thomas: * "A reconnoissance, pushed down to Gadsden to-day, reveals the fact that the rebel Army is not there, and the chances are it has moved west. If it turns up at Guntersville, I will be after it." He writes in his Memoirs:† "There is no doubt that the month of October closed to us looking decidedly squally, but, somehow, I was sustained in the belief that in a very few days the tide would turn." Upon the same page I find the following telegram from General Grant:

"City Point, *November 1st, 1864, 6 p. m.*

"Major General Sherman.

"Do you not think it advisable, now Hood has gone so far north, to entirely ruin him before starting on your proposed campaign? With Hood's Army destroyed, you can go where you please with impunity. I believed and still believe, if you had started south while Hood was in the neighborhood of you, he would have been forced to go after you. Now that he is far away, he might look upon the chase as useless, and he will go on in one direction while you are pushing in another. If you can see a chance of destroying Hood's Army, attend to that first, and make your other move secondary."

General Sherman replied, as follows: ‡

"Rome, Georgia, *November 2d, 1864.*

"Lieutenant General U. S. Grant, City Point, Virginia.

"Your dispatch is received. If I could hope to overhaul Hood, I would turn against him with my whole force; then he would retreat to the southwest, drawing me as a decoy away from Georgia, which is his chief object. If he ventures north of the Tennessee river, I may venture in that direction, and endeavor to get below him on his line of retreat; but thus far he has not gone above the Tennessee river. General Thomas will have a force strong enough to prevent his reaching any country in which we have an interest; and he has orders, if Hood turns

---

* Van Horne's History of the Army of the Cumberland, vol. II, page 181.

† Sherman's Memoirs, vol. II, page 164.

‡ Sherman's Memoirs, vol. II, page 165.

to follow me, to push for Selma, Alabama. No single Army can catch Hood, and I am convinced the best results will follow from our defeating Jeff. Davis's cherished plan of making me leave Georgia by manœuvring. Thus far I have confined my efforts to thwart this plan, and have reduced baggage so that I can pick up and start in any direction; but I regard the pursuit of Hood as useless. Still, if he attempts to invade Middle Tennessee, I will hold Decatur, and be prepared to move in that direction; but, unless I let go Atlanta, my force will not be equal to his.

"W. T. SHERMAN, *Major General.*"

Before my attention was arrested by the above dispatches, I had written those lines which record my surmises in regard to Sherman's and Thomas's movements, during our campaign to the Alabama line. I did not, however, believe that Sherman would follow me to Guntersville, unless I had been able to worst him in battle. No better proof can be adduced of the wisdom of this campaign than the foregoing dispatches, together with our success in drawing Sherman back, within ten days, to Snake Creek Gap, the identical position he occupied in May, 1864. Had the Army been in the fighting condition in which it was at Dalton, or at Franklin, I feel confident of our ability to have at least so crippled the enemy in pitched battle as to have retained possession of the mountains of Georgia. When I consider also the effect of this movement upon the Federal commanders, I cannot but become impressed with the facility with which the Confederate Army would have taken possession of the country as far north as the Ohio, if it had marched in the early Spring of '64, to the rear of the Federals (who were at Chattanooga assembling their forces); and when, in addition to the troops at Dalton, Polk's Army, Longstreet's Corps, and ten thousand men from Beauregard, were proffered for the purpose.

After halting two days at Cross Roads, I decided to make provision for twenty days' supply of rations in the haversacks and wagons; to order a heavy reserve of artillery to accompany the Army, in order to overcome any serious opposition by the Federal gunboats; to cross the Tennessee at or near Guntersville, and again destroy Sherman's communications, at

Stevenson and Bridgeport; to move upon Thomas and Schofield, and attempt to rout and capture their Army before it could reach Nashville. I intended then to march upon that city where I would supply the Army and reinforce it, if possible, by accessions from Tennessee. I was imbued with the belief that I could accomplish this feat, afterward march northeast, pass the Cumberland river at some crossing where the gunboats, if too formidable at other points, were unable to interfere; then move into Kentucky, and take position with our left at or near Richmond, and our right extending toward Hazelgreen, with Pound and Stoney Gaps, in the Cumberland Mountains, at our rear.

In this position I could threaten Cincinnati, and recruit the Army from Kentucky and Tennessee; the former State was reported, at this juncture, to be more aroused and embittered against the Federals than at any period of the war. While Sherman was debating between the alternative of following our Army or marching through Georgia, I hoped, by rapid movements, to achieve these results.

If Sherman cut loose and moved south—as I then believed he would do after I left his front *without previously worsting him in battle*—I would occupy at Richmond, Kentucky, a position of superior advantage, as Sherman, upon his arrival at the sea coast, would be forced to go on board ship, and, after a long detour by water and land, repair to the defence of Kentucky and Ohio or march direct to the support of Grant. If he returned to confront my forces, or followed me directly from Georgia into Tennessee and Kentucky, I hoped then to be in condition to offer battle; and, if blessed with victory, to send reinforcements to General Lee, in Virginia, or to march through the gaps in the Cumberland Mountains, and attack Grant in rear. This latter course I would pursue in the event of defeat or of inability to offer battle to Sherman. If on the other hand he marched to join Grant, I could pass through the Cumberland gaps to Petersburg, and attack Grant in rear, at least two weeks before he, Sherman, could render him assistance. This move, I believed, would defeat Grant, and allow General Lee, in command of our combined Armies, to march upon Washington or turn upon and annihilate Sherman.

Such is the plan which during the 15th and 16th, as we lay in bivouac near Lafayette, I maturely considered, and determined to endeavor to carry out. In accordance therewith, I decided to move to Gadsden, where, if I met General Beauregard, I intended to submit to him the foregoing plan of operations, expressing at the same time my conviction that therein lay the only hope to bring victory to the Confederate arms.

On the 17th, the Army resumed its line of march, and that night camped three miles from the forks of the Alpine, Galesville, and Summerville roads; thence proceeded towards Gadsden.

On the 19th, I sent the following dispatches:

[No. 35.]

"*October 19th.*

"GENERAL BRAGG AND HON. J. A. SEDDON.

"Headquarters will be to-morrow at Gadsden, where I hope not to be delayed more than forty-eight hours, when I shall move for the Tennessee river.

"J. B. HOOD, *General.*"

[No. 36.]

"*October 20th.*

"LIEUTENANT GENERAL TAYLOR, *Mobile.*

"I will move to-morrow for Guntersville on the Tennessee. Please place all the garrison you can at Corinth, and have the railroad iron from there to Memphis taken up as close as possible to Memphis. Have not yet seen General Beauregard. Give me all the assistance you can to get my supplies to Tuscumbia.

"J. B. HOOD, *General.*"

I proposed to move directly on to Guntersville, as indicated to General Taylor, and to take into Tennessee about one-half of Wheeler's cavalry (leaving the remainder to look after Sherman) and to have a depot of supplies at Tuscumbia, in the event I met with defeat in Tennessee.

Shortly after my arrival at Gadsden, General Beauregard reached the same point; I at once unfolded to him my plan, and requested that he confer apart with the corps commanders, Lieutenant Generals Lee and Stewart, and Major General Cheatham. If after calm deliberation, he deemed it expedient we should remain upon the Alabama line and attack Sherman, or take position, entrench, and finally follow on his rear when he moved south, I would of course acquiesce, albeit with reluctance. If, contrariwise, he should agree to my proposed plan to cross into Tennessee, I would move immediately to Guntersville, thence to Stevenson, Bridgeport, and Nashville.

This important question at issue was discussed during the greater part of one night, with maps before us. General Beauregard at length took the ground that if I engaged in the projected campaign, it would be necessary to leave in Georgia all the cavalry at present with the Army, in order to watch and harass Sherman in case he moved south, and to instruct Forrest to join me as soon as I crossed the Tennessee river. To this proposition I acceded. After he had held a separate conference with the corps commanders, we again debated several hours over the course of action to be pursued; and, during the interview, I discovered that he had gone to work in earnest to ascertain, in person, the true condition of the Army; that he had sought information not only from the corps commanders, but from a number of officers, and had reached the same conclusion I had formed at Lafayette: we were not competent to offer pitched battle to Sherman, nor could we follow him south without causing our retrograde movement to be construed by the troops into *a recurrence* of retreat, which would entail desertions, and render the Army of little or no use in its opposition to the enemy's march through Georgia. After two days' deliberation, General Beauregard authorized me, on the evening of the 21st of October, to proceed to the execution of my plan of operations into Tennessee.

At this point, it may be considered, closed the campaign to the Alabama line.

CHAPTER XVI.

TENNESSEE CAMPAIGN—FORREST—WHEELER—DEFLECTION TO FLORENCE—DETENTION—PRESIDENT DAVIS—BEAUREGARD—COLUMBIA—SPRING HILL.

GENERAL BEAUREGARD'S approval of a forward movement into Tennessee was soon made known to the Army. The prospect of again entering that State created great enthusiasm, and from the different encampments arose at intervals that genuine Confederate shout so familiar to every Southern soldier, and which then betokened an improved state of feeling among the troops.

With twenty days' rations in the haversacks and wagons, we marched, on the 22d of October, upon all the roads leading from Gadsden in the direction of Guntersville, on the Tennessee river, and bivouacked that night in the vicinity of Bennetsville.

I here received information that General Forrest was near Jackson, Tennessee, and could not reach the middle portion of this State, as the river was too high. It would, therefore, be impossible for him to join me, if I crossed at Guntersville; as it was regarded as essential that the whole of Wheeler's cavalry remain in Georgia, I decided to deflect westward, effect a junction with Forrest, and then cross the river at Florence. General Beauregard sent orders to him to join me without delay; also dispatched a messenger to hasten forward supplies to Tuscumbia.

The succeeding day, the movement was continued toward Florence, in lieu of Guntersville as I had expected. Lieutenant General Lee's Corps reached the Tennessee, near Florence, on the 30th; Johnson's Division crossed the river, and took possession of that town. My headquarters were during the 27th and 28th at the house of General Garth, near Decatur, where also stopped General Beauregard. While the Army turned Decatur, I ordered a slight demonstration to be made against the town till our forces passed safely beyond, when I moved toward Tuscumbia, at which place I arrived on the 31st of October. Johnson's Division, which held possession of Florence, was reinforced the same day by Clayton's Division.

Thus the Confederate Army rested upon the banks of the Tennessee one month after its departure from Palmetto. It had been almost continuously in motion during this interim; it had by rapid moves and manœuvres, and with only a small loss, drawn Sherman as far north as he stood in the early Spring. The killed and wounded at Allatoona had been replaced by absentees who returned to ranks, and, as usual in such operations, the number of desertions became of no consequence. In addition to the official returns, my authority for the last assertion is Judge Cofer, of Kentucky, who was provost marshal of the Army at this period, and is at present one of the district judges of his State. About two years ago, in Louisville, he informed me that he had been impressed by the small number of desertions reported to him during the campaigns to the rear of Sherman, and into Tennessee.

Notwithstanding my request as early as the 9th of October that the railroad to Decatur be repaired, nothing had been done on the 1st of November towards the accomplishment of this important object, as the following dispatch from the superintendent of the road will show:

"CORINTH, MISSISSIPPI, *November 1st, 1864.*

"GENERAL G. T. BEAUREGARD.

"I fear you have greatly over-estimated the capacity and condition of this railroad to transport the supplies for General Hood's Army.

"Most of the bridges between here and Okolona were destroyed and recently only patched up to pass a few trains of supplies for General Forrest, and are liable to be swept away by freshets which we may soon expect. The cross-ties are so much decayed that three trains ran off yesterday, and the track will be still worse in rainy weather.

"I have called upon General Taylor for additional labor, and will use every effort to forward the supplies, but deem it due to you to advise you of the true condition of the road.

"(Signed) L. J. FLEMING,
"Chief Engineer and General Superintendent M. & O. R. R."

I had expected upon my arrival at Tuscumbia to find additional supplies, and to cross the river at once. Unfortunately, I was constrained to await repairs upon the railroad before a sufficient amount of supplies could be received to sustain the Army till it was able to reach Middle Tennessee.

General Beauregard remained two weeks at Tuscumbia and in its vicinity, during which interval the inaugurated campaign was discussed anew at great length. General Sherman was still in the neighborhood of Rome, and the question arose as to whether we should take trains and return to Georgia to oppose his movements south or endeavor to execute the projected operations into Tennessee and Kentucky. I adhered to the conviction I had held at Lafayette and Gadsden, and a second time desired General Beauregard to consult the corps commanders, together with other officers, in regard to the effect a return to Georgia would produce upon the Army. I also urged the consideration that Thomas would immediately overrun Alabama, if we marched to confront Sherman. I had fixedly determined, unless withheld by Beauregard or the authorities at Richmond, to proceed, as soon as supplies were received, to the execution of the plan submitted at Gadsden.

On the 6th of November, I sent the following dispatch to the President:

[No. 37.]

"HEADQUARTERS TUSCUMBIA,
"*November 6th.*

"HIS EXCELLENCY, PRESIDENT DAVIS, *Richmond.*

"General Wheeler reports from Blue Mountain that Sherman is moving one corps to Tennessee, and three to Marietta. I hope to march for

Middle Tennessee by the eighth or ninth (8th or 9th) inst. Should he move two or three corps south from Atlanta, I think it would be the best thing that could happen for our general good. General Beauregard *agrees with me* as to my plan of operation. Would like to be informed if any forces are sent from Grant or Sheridan, to Nashville.

"J. B. HOOD, *General.*"

At this juncture, I was advised of the President's opposition to the campaign into Tennessee previous to a defeat of Sherman in battle, as is clearly indicated by his reply:

"RICHMOND, *November 7th, 1864.*
"Via Meridian.

"GENERAL J. B. HOOD.

"No troops can have been sent by Grant or Sheridan to Nashville. The latter has attempted to reinforce the former, but Early's movements prevented it. That fact will assure you as to their condition and purposes. The policy of taking advantage of the reported division of his forces, where he cannot re-unite his Army, is too obvious to have been overlooked by you. I therefore take it for granted that you have not been able to avail yourself of that advantage, during his march northward from Atlanta. Hope the opportunity will be offered before he is extensively recruited. If you keep his communications destroyed, he will most probably seek to concentrate for an attack on you. But if, as reported to you, he has sent a large part of his force southward, you may first beat him in detail, and, subsequently, without serious obstruction or danger to the country in your rear, advance to the Ohio river.

"JEFFERSON DAVIS."

The President, as indicated, was evidently under the impression that the Army should have been equal to battle by the time it had reached the Alabama line, and was averse to my going into Tennessee.* He was not, as General Beauregard and myself, acquainted with its true condition. Therefore, a high regard for his views notwithstanding, I continued firm in the belief that the only means to checkmate Sherman, and co-operate with General Lee to save the Confederacy, lay in speedy success in Tennessee and Kentucky, and in my ability finally to attack Grant in rear with my entire force.

* Almost every writer upon the subject of my campaign into Tennessee, has fallen into the popular error that the President ordered me into that State; and, strange to say, General Taylor, brother-in-law of Mr. Davis, has also grossly erred in this regard, when he could have addressed a note to the Chief Executive of the Confederacy and have ascertained the truth.

On the 9th, I telegraphed to the Secretary of War:

[No. 38.]

"HEADQUARTERS TUSCUMBIA,
"*November 9th.*

"HON. J. A. SEDDON, *Richmond, Va.*

"Information received places Sherman's Army as follows: One corps at Atlanta, two corps at or near Marietta; and three at or north of Chattanooga. Heavy rains will delay the operations of this Army a few days.

"J. B. HOOD, *General.*"

Although every possible effort was made to expedite the repairs upon the railroad, the work progressed slowly. Heavy rains in that section of the country also interfered with the completion of the road.

I informed General Beauregard of the President's opposition to my plan, and, on the 12th, replied to His Excellency, as follows:

[No. 39.]

"HEADQUARTERS NEAR FLORENCE, ALABAMA,
"*November 12th, 1864.*

"HIS EXCELLENCY, THE PRESIDENT, *Richmond, Virginia.*

"Your telegram of the 7th received to-day. When I moved out from Atlanta, he (Sherman) came with five corps, and kept them united until I moved from Gadsden to this point, entrenching himself wherever he halted. It was only after I reached this point that he divided his force. After my descent upon the railroad and upon Dalton, I did not regard this Army in proper condition for a pitched battle. It is now in excellent spirits, and confident. Before leaving Gadsden, I urged on General Beauregard to send General Forrest across the Tennessee river. This he ordered; and I intended, when leaving Gadsden, to cross the river at or near Gunter's Landing. Finding, however, when I reached that vicinity, that Forrest had not crossed, I could not, without his co-operation, pass the river there, as I required Wheeler to look after my right flank. Forrest has not yet crossed over, but is moving upon this side of the river, and will join me here.

"This circumstance, high water, and the fact that I had to draw supplies from and through a department not under my command—involving delay in their reaching me—have retarded my operations. As soon as Forrest joins me, which will be in a few days, I shall be able to move forward. Without the assistance of Forrest, I cannot secure my wagon trains when across the river. You may rely upon my striking the enemy whenever a suitable opportunity presents itself, and that I will spare no effort to make that opportunity.

"J. B. HOOD, *General.*"

On the 13th, I established my headquarters in Florence, upon the north branch of the Tennessee, and the following day General Forrest, with his command, reported for duty. On the 15th, the remainder of Lee's Corps crossed the river, and bivouacked in advance of Florence. Stewart's and Cheatham's Corps were instructed also to cross the same day.

Upon this date, I received the following from General Beauregard:

"HEADQUARTERS MILITARY DIVISION OF THE WEST,
"TUSCUMBIA, ALABAMA, *November 15th, 1864.*

"GENERAL:—As you seemed on yesterday to have misunderstood my verbal communication of the 13th inst., through my chief of staff, I deem it of sufficient importance to communicate in writing, what I had instructed him to say relative to the movement of the Army of Tennessee.

"I instructed him to tell you that in consequence of the information received the night previous, to-wit, the apparent confirmation of the concentration of the bulk of Sherman's Army in Middle Tennessee (at Pulaski, Huntsville, and Decatur), the arrival of Canty and part of his forces at Memphis, and the condition of Cobb's and Smith's forces at Lovejoy's Station, I desired to confer further with you before you commenced the projected movement into Middle Tennessee, now partly in process of execution; that is, Lee's Corps already in advance of Florence, and Stewart's and Cheatham's Corps under orders to cross the river.

"My purpose was to call again your attention as I did yesterday:

"1st. To the necessity of guarding well your left flank, and rear, in advancing towards Lawrenceburg and Pulaski, against a sudden offensive movement of the enemy from Huntsville or Athens, across the Elk river.

"2d. To securing against the passage of the enemy's gunboats another point (about Savannah or Clifton) besides Florence for the Army to recross the Tennessee, in the event of disaster.

"3d. To giving still greater protection to Corinth, and the M. and O. R. R. to that point.

"I was aware that these points had already been discussed between us, but my anxiety for the safety of the troops under your command, made it incumbent on me to call again your attention to these important matters.

"I wish also to inform you that the third point mentioned may require greater time than was at first supposed necessary. All orders for completing the defences of Corinth, repairing and prosecuting vigorously the work on the M. and C. R. R. to this place, and for repairing the

M. and O. R. R. from Okolona to Bethel, have been given, and are being carried out as rapidly as the limited means of the engineer and quarter master's departments will permit. It is at present reported that the railroads referred to will be completed in from fifteen to twenty days; but it is not unreasonable to suppose that the prevailing unfavorable weather will delay the work one or two weeks longer.

"General Taylor and myself will always be anxious to aid you in your present campaign, with all the means at our control; but, these being limited, ample previous notice of what may be required, should be given to enable us to make all necessary preparations.

"It will also give me pleasure to confer on you such powers as you may deem necessary to secure your communications, repair roads, and hasten supplies to your Army, whilst operating in the department of General Taylor.

"Respectfully, your obedient servant,

"G. T. BEAUREGARD, *General.*"

About the time all necessary preparations verged to a completion, and I anticipated to move forward once more, heavy rains again delayed our supplies, as will be shown by the subjoined communication from Colonel Brent:

"HEADQUARTERS MILITARY DIVISION OF THE WEST,
"TUSCUMBIA, ALABAMA, *November 17th, 1864.*

"GENERAL:—General Beauregard instructs me to say that a bridge about three miles from Tuscumbia on road to Cherokee, is now being constructed, and that, for want of workmen, it cannot be completed in less than five or six days, and at this point the road is almost impassable. There are also other points on the road which will become impassable, should the rain continue. He thinks it important that a proper force should be sent to complete the improvements as early as possible.

"Respectfully, your obedient servant,

"GEORGE W. BRENT,
"Colonel and A. A. G."

In compliance with this request, working parties were at once detailed, and sent to different points on the railroad; wagons were also dispatched to aid in the transportation of supplies. The officer in charge was instructed to require the men to labor unceasingly toward the accomplishment of this important object.

On the 17th, General Beauregard issued the following order previous to his departure for Montgomery, Alabama:

"HEADQUARTERS MILITARY DIVISION OF THE WEST,
"TUSCUMBIA, ALABAMA, *November 17th, 1864.*

"GENERAL:—General Beauregard desires me to say that he desires you will take the offensive at the earliest practicable moment, striking the enemy whilst thus dispersed, and by this means distract Sherman's advance into Georgia.

"To relieve you from any embarrassment whilst operating in North Alabama and Middle Tennessee, he authorizes you to issue all such orders, in General Taylor's Department, you may deem necessary to secure the efficient and successful administration and operation of your Army—sending General Taylor copies of all orders.

"He wishes you to send forthwith to Major General Wheeler one brigade of cavalry of Jackson's Division, and the balance of that Division as soon as it can be spared, should Sherman advance into Georgia; and also to advise General Wheeler that in such case Clanton's brigade is subject to his orders.

"The headquarters of this military division will be removed, in the morning, from this place to Montgomery, Alabama.

"I am, General, respectfully, your obedient servant,

"GEORGE W. BRENT,
"Colonel and A. A. G.

"General J. B. HOOD,
"Commanding Army of Tennessee."

The ensuing day, I replied:

[No. 537.]

"FLORENCE, ALABAMA, *November 18th, 1864.*

"I will send two batteries from the Army to Corinth. General Forrest thinks his force of cavalry entirely insufficient without Jackson's Division.

"J. B. HOOD, *General.*"

The working parties on the railroad having succeeded in pushing forward the supplies, I also telegraphed to him, on the 19th, that I would resume the line of march at the earliest practicable moment.

Information had, in the meantime, reached me that Sherman was advancing south, from Atlanta. He marched out of that fated city on the 16th, and thus describes his going forth: *

---

* Sherman's Memoirs, vol. II, page 178.

"About 7 a. m., of November 16th, we rode out of Atlanta by the Decatur road, filled by the marching troops and wagons of the Fourteenth Corps; and reaching the hill, just outside of the old rebel works, we naturally paused to look back upon the scenes of our past battles. We stood upon the very ground whereon was fought the bloody battle of July 22d, and could see the copse of wood where McPherson fell. Behind us lay Atlanta, smouldering and in ruins, the black smoke rising high in air, and hanging like a pall over the ruined city."

Thus were two opposing Armies destined to move in opposite directions, each hoping to achieve glorious results.

I well knew the delay at Tuscumbia would accrue to the advantage of Sherman, as he would thereby be allowed time to repair his railroad, and at least start to the rear all surplus material. I believed, however, I could still get between Thomas's forces and Nashville, and rout them; furthermore, effect such manœuvres as to insure to our troops an easy victory. These convictions counterbalanced my regret that Sherman was permitted to traverse Georgia "unopposed," as he himself admits.

General Beauregard had moved in the direction of Georgia to assemble all available forces to oppose Sherman's advance. At the time I made my official report, I was furnished a copy of his letter to President Davis, stating in full the reasons which had induced him to approve my campaign, and enumerating the difficulties, at this crisis, to be encountered in a movement southward to Georgia. This letter is dated the 6th of December, but I insert it at this point, since it treats of events under consideration, and which occurred just prior to the advance into Tennessee:

"AUGUSTA, GEORGIA, *December 6th, 1864.*

"TO HIS EXCELLENCY, JEFFERSON DAVIS, *President Confederate States.*

"SIR:—Your letter of the 30th ult., acknowledging the receipt of my telegram of the 24th November, was received by me on the road from Macon to this place.

"With the limited reliable means at our command, I believe that all that could be, has been done, under existing circumstances to oppose the advance of Sherman's forces towards the Atlantic coast. That we have not thus far been more successful, none can regret more than myself; but he will doubtless be prevented from capturing Augusta, Charleston, and Savannah, and he may yet be made to experience serious loss before reaching the coast.

"On the 16th of November, when about leaving Tuscumbia, Alabama, on a tour of inspection to Corinth, Mississippi, I was informed by General Hood of the report just received by him, that Sherman would probably move from Atlanta into Georgia. I instructed him at once to repeat his orders to General Wheeler to watch closely Sherman's movements, and, should he move as reported, to attack and harass him at all favorable points.

"I telegraphed to Lieutenant General Taylor at Selma, Alabama, to call on Governor Watts, of Alabama, and Governor Clarke, of Mississippi, for all the State troops that they could furnish; and with all the available moveable forces of his department, to keep himself in readiness to move at a moment's notice, to the assistance of Major General Howell Cobb and Major General G. W. Smith, who were then at or about Griffin, Georgia, threatening Atlanta.

"I also telegraphed to General Cobb to call upon Governor Brown, of Georgia, and Governor Bonham, of South Carolina, for all the State troops that could be collected.

"I made all necessary preparations to repair forthwith to Georgia, in the event of Sherman's executing his reported movement.

"On my arrival at Corinth, on the 18th of November, having been informed that Sherman had commenced his movement, I issued all necessary orders to meet the emergency, including an order to General Hood to send one division of cavalry (Jackson's) to reinforce Wheeler; but this order was suspended by him, his objection being that his cavalry could not be reduced without endangering the success of his campaign in Tennessee, and that General Wheeler had already thirteen brigades under his command. I finally instructed him to send only *one* brigade, if he contemplated taking the offensive at once, as had already been decided upon. I then left Corinth for Macon, where I arrived on the 24th of November.

"I did not countermand the campaign in Tennessee to pursue Sherman with Hood's Army for the following reasons:

"1st. The roads and creeks from the Tennessee to the Coosa river across Sand and Lookout Mountains had been, by the prevailing heavy rains, rendered almost impassable to artillery and wagon trains.

"2d. General Sherman, with an Army better appointed, had already the start about two hundred and seventy-five miles on comparatively good roads. The transfer of Hood's Army into Georgia could not have been more expeditious by railway than by marching through the country, on account of the delays unavoidably resulting from the condition of the railroads.

"3d. To pursue Sherman, the passage of the Army of Tennessee would, necessarily, have been over roads with all the bridges destroyed, and through a devastated country, affording no subsistence or forage; and, moreover, it was feared that a retrograde movement on our part would seriously deplete the Army by desertions.

"4th. To have sent off the most or the whole of the Army of Tennessee in pursuit of Sherman, would have opened to Thomas's force the richest portion of the State of Alabama, and would have made nearly certain the capture of Montgomery, Selma, and Mobile, without insuring the defeat of Sherman.

"5th. In October last, when passing through Georgia to assume command of the Military Division of the West, I was informed by Governor Brown that he could probably raise, in case of necessity, about six thousand (6000) men, which I supposed might be doubled in a levy '*en masse*.'

"General Cobb informed me, at the same time, that at Augusta, Macon, and Columbus, he had about six thousand five hundred (6500) local troops, and that he hoped shortly to have collected at his reserve and convalescent camps, near Macon, twenty-five hundred (2500) men. Of these nine thousand (9000) men, he supposed about one-half, or five thousand (5000), could be made available as moveable troops for an emergency.

"To oppose the advance of the enemy from Atlanta, the State of Georgia would thus have probably nineteen thousand (19,000) men, to which number must be added the thirteen brigades of Wheeler's cavalry, amounting to about seven thousand (7000) men. The troops which would have been collected from Savannah, South Carolina, and North Carolina, before Sherman's forces could reach the Atlantic coast, would have amounted, it was supposed, to about five thousand (5000) men.

"Thus it was a reasonable supposition that about twenty-nine or thirty thousand (29,000 or 30,000) men could be collected in time to defend the State of Georgia, and insure the destruction of Sherman's Army, estimated by me at about thirty-six thousand (36,000) effectives of all arms—their cavalry, about four thousand (4000), being included in the estimate.

"Under these circumstances, after consultation with General Hood, I concluded to allow him to prosecute with vigor his campaign into Tennessee and Kentucky, hoping that by defeating Thomas's Army and such other forces as might hastily be sent against him, he would compel Sherman, should he reach the coast of Georgia or South Carolina, to repair at once to the defence of Kentucky and, perhaps, Ohio, and thus prevent him from reinforcing Grant. Meanwhile, supplies might be sent to Virginia from Middle and East Tennessee, thus relieving Georgia from the present constant drain upon its limited resources.

"I remain very respectfully, your obedient servant.

"G. T. BEAUREGARD, *General*."

The writer of the above letter greatly over-estimated the number of troops which he hoped to collect for the defence of Georgia. He published a stirring appeal to the people of the State to rally and drive back the enemy, but he was not successful in obtaining even one-half the number of men he anticipated, and a great portion of those who responded to his call were irregular troops. The Honorable B. H. Hill, in an eloquent address, also urged the people to action, but, as I have already stated, the country at this period was well nigh drained of all its resources.

General Beauregard, as previously mentioned, left me on the 17th of November. On the 19th, the preliminaries to the campaign being completed, the cavalry was ordered to move forward. The succeeding day, Lee's Corps marched to the front a distance of about ten miles on the Chisholm road, between the Lawrenceburg and Waynesboro' roads.

The same day, I received the following dispatch from General Beauregard:

"WEST POINT, *20th, 10 a. m.*

"GENERAL J. B. HOOD.

"Push on active offensive immediately. Colonel Brent informs me first order for movement of one of Jackson's brigades to Wheeler has been suspended by you. It is indispensable it should be sent by best and quickest route to Newnan to cut off communications of enemy with Kingston, and to protect (here in cipher, of which I have not the key). I have appealed to the people of Georgia to defend their homes.

"G. T. BEAUREGARD, *General.*"

On the 20th of November, Stewart's Corps having crossed the Tennessee and bivouacked several miles beyond on the Lawrenceburg road, orders were issued that the entire Army move at an early hour the next morning. Lee's and Stewart's Corps marched upon the Chisholm and the Lawrenceburg roads, and Cheatham's Corps upon the Waynesboro' road.

Early dawn of the 21st found the Army in motion. I hoped by a rapid march to get in rear of Schofield's forces, then at Pulaski, before they were able to reach Duck river. That night headquarters were established at Rawhide, twelve miles north of Florence, on the Waynesboro' road.

The march was resumed on the 22d, and continued till the 27th, upon which date the troops, having taken advantage of every available road, reached Columbia, via Mount Pleasant. Forrest operated in our front against the enemy's cavalry which he easily drove from one position to another.

The Federals at Pulaski became alarmed, and, by forced marches day and night, reached Columbia, upon Duck river, just in time to prevent our troops from cutting them off. Van Horne, in his History of the Army of the Cumberland, thus mentions their narrow escape: *

> "General Hood's rapid advance had been made with the hope of cutting off General Schofield from Columbia, and barely failed in this object, as the National troops gained the place by a night march."

The enemy having formed line of battle around Columbia, Lee's Corps filed into position with its right upon the Mount Pleasant pike; Stewart's formed on Lee's right, his own right flank extending to the Pulaski pike; and Cheatham established his left on the latter pike, with his right resting on Duck river. Army headquarters were established at the residence of Mrs. Warfield, about three miles south of Columbia.

The two Armies lay opposite each other during the 27th. The Federals being entrenched, I determined not to attack them in their breastworks, if I could possibly avoid it, but to permit them to cross undisturbed to the north bank of Duck river that night, as I supposed they would do; to hasten preparations, and endeavor to place the main body of the Confederate Army at Spring Hill, twelve miles directly in the enemy's rear, and about mid-way upon the only pike leading to Franklin; to attack as the Federals retreated, and put to rout and capture, if possible, their Army which was the sole obstacle between our forces and Nashville—in truth, the only barrier to the success of the campaign.

---

*Van Horne's Army of the Cumberland, vol. II, page 189.

I was confident that after Schofield had crossed the river and placed that obstruction between our respective Armies, he would feel in security, and would remain in his position at least a sufficient length of time to allow me to throw pontoons across the river about three miles above his left flank, and, by a bold and rapid march together with heavy demonstrations in his front, gain his rear before he was fully apprised of my object.

The situation presented an occasion for one of those interesting and beautiful moves upon the chess-board of war, to perform which I had often desired an opportunity. As stated in a letter to General Longstreet, I urgently appealed for authority to turn the Federal left at Round Top Mountain. I had beheld with admiration the noble deeds and grand results achieved by the immortal Jackson in similar manœuvres; I had seen his Corps made equal to ten times its number by a sudden attack on the enemy's rear, and I hoped in this instance to be able to profit by the teaching of my illustrious countryman. As I apprehended unnecessary and fatal delay might be occasioned by the appearance of the enemy on the line of march to the rear, I decided to bridge the river that night, and move at dawn the next morning with Cheatham's Corps—whose right was then resting near the point selected for a crossing—together with Stewart's Corps and Johnston's Division, of Lee's Corps, and to leave Lieutenant General Lee with Stevenson's and Clayton's Divisions and the bulk of the artillery, to demonstrate heavily against Schofield, and follow him if he retired.

Since I had attempted this same movement on the 22d of July, and had been unable to secure its success, I resolved to go in person at the head of the advance brigade, and lead the Army to Spring Hill.

Colonel Prestman and his assistants laid the pontoons during the night of the 28th, about three miles above Columbia; orders to move at dawn the following day having been issued to the two corps and the division above mentioned, I rode with my staff to Cheatham's right, passed over the bridge soon after daybreak, and moved forward at the head of Granberry's Texas brigade, of Cleburne's Division, with instructions that the remaining corps and divisions follow, and at the same time keep well closed up during the march.

General Forrest had crossed the evening previous and moved to the front and right. I threw forward a few skirmishers who advanced at as rapid a pace as I supposed the troops could possibly proceed.

During the march, the Federal cavalry appeared on the hills to our left; not a moment, however, was lost on that account, as the Army was marching by the right flank and was prepared to face at any instant in their direction. No attention, therefore, was paid to the enemy, save to throw out a few sharp-shooters in his front. I well knew that to stop and lose time in reconnoitering would defeat my object, which was to reach the enemy's rear and cut him off from Nashville.

I also knew that Schofield was occupied in his front, since I could distinctly hear the roar of Lee's artillery at Columbia, whilst a feint was made to cross the river.

Thus I led the main body of the Army to within about two miles and in full view of the pike from Columbia to Spring Hill and Franklin. I here halted about 3 p. m., and requested General Cheatham, commanding the leading corps, and Major General Cleburne to advance to the spot where, sitting upon my horse, I had in sight the enemy's wagons and men passing at double-quick along the Franklin pike. As these officers approached, I spoke to Cheatham in the following words which I quote almost verbatim, as they have remained indelibly engraved upon my memory ever since that fatal day: "General, do you see the enemy there, retreating rapidly to escape us?" He answered in the affirmative. "Go," I continued, "with your Corps, take possession of and hold that pike at or near Spring Hill. Accept whatever comes, and turn all those wagons over to our side of the house." Then addressing Cleburne, I said, "General, you have heard the orders just given. You have one of my best divisions. Go with General Cheatham, assist him in every way you can, and do as he directs." Again, as a parting injunction to them, I added, "Go and do this at once. Stewart is near at hand, and I will have him double-quick his men to the front."

They immediately sent staff officers to hurry the men forward, and moved off with their troops at a quick pace in the direction of the enemy. I dispatched several of my staff to the rear, with orders to Stewart and Johnson to make all possible haste. Meantime I rode to one side, and looked on at Cleburne's Division, followed by the remainder of Cheatham's Corps, as it marched by seemingly ready for battle.

Within about one-half hour from the time Cheatham left me, skirmishing began with the enemy, when I rode forward to a point nearer the pike, and again sent a staff officer to Stewart and Johnson to push forward. At the same time, I dispatched a messenger to General Cheatham to lose no time in gaining possession of the pike at Spring Hill. It was reported back that he was about to do so.

Listening attentively to the fire of the skirmishers in that direction, I discovered there was no continued roar of musketry, and being aware of the quick approach of darkness, after four o'clock at that season of the year, I became somewhat uneasy, and again ordered an officer to go to General Cheatham, inform him that his supports were very near at hand, that he must attack at once, if he had not already so done, and take and hold possession of the pike. Shortly afterwards, I entrusted another officer with the same message, and, if my memory is not treacherous, finally requested the Governor of Tennessee, Isham G. Harris, to hasten forward and impress upon Cheatham the importance of action without delay. I knew no large force of the enemy could be at Spring Hill, as couriers reported Schofield's main body still in front of Lee, at Columbia, up to a late hour in the day. I thought it probable that Cheatham had taken possession of Spring Hill without encountering material opposition, or had formed line across the pike, north of the town, and entrenched without coming in serious contact with the enemy, which would account for the little musketry heard in his direction. However, to ascertain the truth, I sent an officer to ask Cheatham if he held the pike, and to inform him of the arrival of Stewart, whose Corps I intended to throw on his left, in order to assail the Federals in flank that evening or the next morning, as they approached and formed to attack Cheatham. At this juncture, the last messenger returned with the report that the road had not been taken possession of. General Stewart was then ordered to proceed to the right of Cheatham and place his Corps across the pike, north of Spring Hill.

By this hour, however, twilight was upon us, when General Cheatham rode up in person. I at once directed Stewart to halt, and, turning to Cheatham, I exclaimed with deep emotion, as I felt the golden opportunity fast slipping from me, "General, why in the name of God have you not attacked the enemy, and taken possession of that pike?" He replied that the line looked a little too long for him, and that Stewart should first form on his right. I could hardly believe it possible that this brave old soldier, who had given proof of such courage and ability upon so many hard-fought fields, would even make such a report. After leading him within full view of the enemy, and pointing out to him the Federals, retreating in great haste and confusion, along the pike, and then giving explicit orders to attack, I would as soon have expected midday to turn into darkness as for him to have disobeyed my orders. I then asked General Cheatham whether or not Stewart's Corps, if formed on the right, would extend across the pike. He answered in the affirmative. Guides were at once furnished to point out Cheatham's right to General Stewart, who was ordered to form thereon, with his right extending across the pike. Darkness, however, which was increased by large shade trees in that vicinity, soon closed upon us, and Stewart's Corps, after much annoyance, went into bivouac for the night, near but not across the pike, at about eleven or twelve o'clock.

It was reported to me after this hour that the enemy was marching along the road, almost under the light of the camp-fires of the main body of the Army. I sent anew to General Cheatham to know if at least a line of skirmishers could not be advanced, in order to throw the Federals in confusion, to delay their march, and allow us a chance to attack in the morning. Nothing was done. The Federals, with immense wagon trains, were permitted to march by us the remainder of the night, within gunshot of our lines. I could not succeed in arousing the troops to action, when one good division would have sufficed to do the work. One good division, I re-assert, could have routed that portion of the enemy which was at Spring Hill; have taken possession of and formed line across the road; and thus have made it an easy matter to Stewart's Corps, Johnston's Division, and Lee's two Divisions from Columbia, to have enveloped, routed, and captured Schofield's Army that afternoon and the ensuing day. General Forrest gallantly opposed the enemy further down to our right to the full extent of his power; beyond this effort, nothing whatever was done, although never was a grander opportunity offered to utterly rout and destroy the Federal Army.

Had I dreamed one moment that Cheatham would have failed to give battle, or at least to take position across the pike and force the enemy to assault him, I would have ridden, myself, to the front, and led the troops into action. Although it is right and proper that a Commander-in-Chief, in the event of disaster to a portion of his line during an engagement, should endeavor in person to rally the troops, it is not expected nor considered expedient that he should inaugurate a battle by leading a division or brigade. Had I done so, my opponents would have just cause for the charge of recklessness. I would, nevertheless, have risked my life in this instance, had I conceived the possibility of the disregard of my orders, on the part of this officer. General Lee was in a measure thwarted by the same want of prompt action, at Gettysburg. Whilst I failed utterly to bring on battle at Spring Hill, he was unable to get the corps of his Army to attack and co-operate, as desired. He was thus checkmated for two days, and finally lost the battle. Had our immortal Chieftain foreseen the result of this inactivity, he would, doubtless, have ordered and acted differently.

Before proceeding further, I will produce additional evidence from Federal sources, in order to make still more manifest the opportunity which was lost to the Confederate arms on the 29th of November, at Spring Hill.

Shortly after the war, I met in New Orleans Colonel Fullerton, of the United States Army; he was Schofield's adjutant general at the time of these events, in connection with which he wrote me the following:

"NEW ORLEANS, LA., *October 20th, 1865.*

"TO GENERAL HOOD.

"GENERAL:—The only body of United States troops on the battlefield of Spring Hill, Tennessee, on the 29th of November, 1864, was the Second Division of the Fourth Army Corps. I think the division was less than four thousand (4000) strong. There were no other United States troops in or about Spring Hill on that day but one or two hundred cavalrymen and perhaps fifty or sixty infantrymen (post troops). The rest of General Schofield's Army was in the vicinity of Columbia, on the north side of Duck river, and none of these troops began to arrive at Spring Hill until after 9 p. m. I arrived in Spring Hill with the Second Division of the Fourth Corps, and remained there till nearly daylight when I went to Franklin with the rear of the Army. I was at the time lieutenant colonel and assistant adjutant general of the Fourth Army Corps.

"J. S. FULLERTON,

"Brevet Brigadier General, United States Volunteers."

Van Horne, in his History of the Army of the Cumberland, informs us that at 3 p. m., when the Confederate Army was already at Spring Hill, the Federal commander became apprised of our move in his rear, and thus describes his retreat: *

"His (Lee's) repeated attacks were all repulsed by General Cox, and at 3 p. m., General Schofield became satisfied that the enemy would not attack on Duck river, but was moving two corps directly on Spring Hill. He then gave orders for the withdrawal. * * *

* Vol. II, page 194.

"There was some delay at Rutherford's creek, as the bridge was inadequate for the emergency, but nevertheless the divisions, one after another, arrived at Spring Hill—the foremost of the three at 11 p. m. The enemy's pickets fired into the column frequently, but as they did not come upon the road, the National troops gave no response. The enemy were so close to the road, that when a column was not moving upon it, it was difficult for a single horseman to pass." *

"There was momentary expectation that this great Army would take a step forward, and press troops, artillery, and trains from the road in confusion and rout; but still the movement went on without interruption by the enemy."†

"Rarely has an Army escaped so easily from a peril so threatening."‡

In connection with this grave misfortune, I must here record an act of candor and nobility upon the part of General Cheatham, which proves him to be equally generous-hearted and brave. I was, necessarily, much pained by the disappointment suffered, and, a few days later, telegraphed to Richmond, to withdraw my previous recommendation for his promotion, and to request that another be assigned to the command of his Corps. Before the receipt of a reply, this officer called at my headquarters—then at the residence of Mr. Overton, six miles from Nashville—and, standing in my presence, spoke an honest avowal of his error, in the acknowledgment that he felt we had lost a brilliant opportunity at Spring Hill to deal the enemy a crushing blow, and that he was greatly to blame. I telegraphed and wrote to the War Department to withdraw my application for his removal, in the belief that, inspired with an ambition to retrieve his short-coming, he would prove in the future doubly zealous in the service of his country.

The following are the dispatches above referred to:

"HEADQUARTERS, SIX MILES FROM NASHVILLE,
"ON FRANKLIN PIKE, *December 7th, 1864.*

"HONORABLE J. A. SEDDON.

"I withdraw my *recommendation*§ in favor of the *promotion* of *Major General Cheatham* for reasons which I will write more fully.

"J. B. HOOD, *General.*"

---

* Vol. II, page 195. † Vol. II, pages 194, 195.

‡ Van Horne's A. C., vol. II, page 196.

§ The words in italics were in cypher.

"HEADQUARTERS, SIX MILES FROM NASHVILLE,
"ON FRANKLIN PIKE, *December 8th, 1864.*

"HONORABLE J. A. SEDDON, *Secretary of War.*
"GENERAL G. T. BEAUREGARD, *Macon, Ga.*

"A good *Lieutenant General* should be sent here at once to *command* the corps *now commanded* by *Major General Cheatham.* I have no one to recommend for the position.

"J. B. HOOD, *General.*"

"HEADQUARTERS, SIX MILES FROM NASHVILLE,
"ON FRANKLIN PIKE, *December 8th, 1864.*

"HONORABLE J. A. SEDDON.

"*Major General Cheatham* made a failure on the 30th of November, which *will be a lesson to him.* I think it best he should remain in his position for the present. I withdraw my telegrams of yesterday and to-day on this subject.

"J. B. HOOD, *General.*"

On the 11th of December I wrote the Hon. Mr. Seddon:*

* * * * "Major General Cheatham has frankly confessed the great error of which he was guilty, and attaches much blame to himself. While his error lost so much to the country, it has been a severe lesson to him, by which he will profit in the future. In consideration of this, and of his previous conduct, I think that it is best that he should retain, for the present, the command he now holds." * * * * * * * *

The best move in my career as a soldier, I was thus destined to behold come to naught. The discovery that the Army, after a forward march of one hundred and eighty miles, was still, seemingly, unwilling to accept battle unless under the protection of breastworks, caused me to experience grave concern. In my inmost heart I questioned whether or not I would ever succeed in eradicating this evil. It seemed to me I had exhausted every means in the power of one man to remove this stumbling block to the Army of Tennessee. And I will here inquire, in vindication of its fair name, if any intelligent man of that Army supposes one moment that these same troops, one year previous, would, even without orders to

---

*See letter and telegram to Secretary of War, Appendix, page 356.

attack, have allowed the enemy to pass them at Rocky-faced Ridge, as he did at Spring Hill.

Lieutenant General Lee performed his duty, at Columbia, with great skill and fidelity which were crowned with entire success: he attained the object of the demonstration, which was to keep the Federals in ignorance of our movements till sufficient time had been allowed the Army to reach the desired point. Colonel Beckham, chief of artillery in Lee's Corps, and one of the most promising officers of his rank, was unfortunately killed on the 29th, during the heavy cannonade in front of that town. On the morning of the 30th of November, Lee was on the march up the Franklin pike, when the main body of the Army, at Spring Hill, awoke to find the Federals had disappeared.

I hereupon decided, before the enemy would be able to reach his stronghold at Nashville, to make that same afternoon another and final effort to overtake and rout him, and drive him in the Big Harpeth river at Franklin, since I could no longer hope to get between him and Nashville, by reason of the short distance from Franklin to that city, and the advantage which the Federals enjoyed in the possession of the direct road.

CHAPTER XVII.

TENNESSEE CAMPAIGN — FRANKLIN — NASHVILLE — RETREAT — TUPELO—RETURN TO RICHMOND—SURRENDER AT NATCHEZ, MISSISSIPPI.

At early dawn the troops were put in motion in the direction of Franklin, marching as rapidly as possible to overtake the enemy before he crossed the Big Harpeth, eighteen miles from Spring Hill. Lieutenant General Lee had crossed Duck river after dark the night previous, and, in order to reach Franklin, was obliged to march a distance of thirty miles. The head of his column arrived at Spring Hill at 9 a. m. on the 30th, and, after a short rest, followed in the wake of the main body.

A sudden change in sentiment here took place among officers and men: the Army became metamorphosed, as it were, in one night. A general feeling of mortification and disappointment pervaded its ranks. The troops appeared to recognize that a rare opportunity had been totally disregarded, and manifested, seemingly, a determination to retrieve, if possible, the fearful blunder of the previous afternoon and night. The feeling existed which sometimes induces men who have long been wedded to but one policy to look beyond the sphere of their own convictions, and, at least, be willing to make trial of another course of action.

Stewart's Corps was first in order of march; Cheatham followed immediately, and Lieutenant General Lee in rear.

Within about three miles of Franklin, the enemy was discevered on the ridge over which passes the turnpike. As soon as the Confederate troops began to deploy, and skirmishers were thrown forward, the Federals withdrew slowly to the environs of the town.

It was about 3 p. m. when Lieutenant General Stewart moved to the right of the pike and began to establish his position in front of the enemy. Major General Cheatham's Corps, as it arrived in turn, filed off to the left of the road, and was also disposed in line of battle. The artillery was instructed to take no part in the engagement, on account of the danger to which women and children in the village would be exposed. General Forrest was ordered to post cavalry on both flanks, and, if the assault proved successful, to complete the ruin of the enemy by capturing those who attempted to escape in the direction of Nashville. Lee's Corps, as it arrived, was held in reserve, owing to the lateness of the hour and my inability, consequently, to post it on the extreme left. Schofield's position was rendered favorable for defence by open ground in front, and temporary entrenchments which the Federals had had time to throw up, notwithstanding the Confederate forces had marched in pursuit with all possible speed. At one or two points, along a short space, a slight abatis had been hastily constructed, by felling some small locust saplings in the vicinity.

Soon after Cheatham's Corps was massed on the left, Major General Cleburne came to me where I was seated on my horse in rear of the line, and asked permission to form his Division in two, or, if I remember correctly, three lines for the assault. I at once granted his request, stating that I desired the Federals to be driven into the river in their immediate rear and directing him to advise me as soon as he had completed the new disposition of his troops. Shortly afterward, Cheatham and Stewart reported all in readiness for action, and received orders to drive the enemy from his position into the river *at all hazards*. About that time Cleburne returned, and, expressing himself with an enthusiasm which he had never before betrayed in our intercourse, said, "General, I am ready, and have more hope in the final success of our cause than I have had at any time since the first gun was fired." I replied, "God grant it!" He turned and moved at once toward the head of his Division; a few moments thereafter, he was lost to my sight in the tumult of battle. These last words, spoken to me by this brave and distinguished soldier, I have often recalled; they can never leave my memory, as within forty minutes after he had uttered them, he lay lifeless upon or near the breastworks of the foe.

The two corps advanced in battle array at about 4 p. m., and soon swept away the first line of the Federals, who were driven back upon the main line. At this moment, resounded a concentrated roar of musketry, which recalled to me some of the deadliest struggles in Virginia, and which now proclaimed that the possession of Nashville was once more dependent upon the fortunes of war. The conflict continued to rage with intense fury; our troops succeeded in breaking the main line at one or more points, capturing and turning some of the guns on their opponents.

Just at this critical moment of the battle, a brigade of the enemy, reported to have been Stanley's, gallantly charged, and restored the Federal line, capturing at the same time about one thousand of our troops within the entrenchments. Still the ground was obstinately contested, and, at several points upon the immediate sides of the breastworks, the combatants endeavored to use the musket upon one another, by inverting and raising it perpendicularly, in order to fire; neither antagonist, at this juncture, was able to retreat without almost a certainty of death. It was reported that soldiers were even dragged from one side of the breastworks to the other by men reaching over hurriedly and seizing their enemy by the hair or the collar.

Just before dark Johnston's Division, of Lee's Corps, moved gallantly to the support of Cheatham; although it made a desperate charge and succeeded in capturing three stands of colors, it did not effect a permanent breach in the line of the enemy. The two remaining divisions could not unfortunately become engaged owing to the obscurity of night.* The struggle continued with more or less violence until 9 p. m., when followed skirmishing and much desultory firing until about 3 a. m. the ensuing morning. The enemy then withdrew, leaving his dead and wounded upon the field. Thus terminated one of the fiercest conflicts of the war.

Nightfall which closed in upon us so soon after the inauguration of the battle prevented the formation and participation of Lee's entire Corps on the extreme left. This, it may safely be asserted, saved Schofield's Army from destruction. I might, with equal assurance, assert that had Lieutenant General Lee been in advance at Spring Hill the previous afternoon, Schofield's Army never would have passed that point.

Shortly afterward I sent the following dispatch to the Secretary of War and to General Beauregard:

[No. 541.]

"HEADQUARTERS, SIX MILES TO NASHVILLE,
"*December 3d.*

"About 4 p. m., November 30th, we attacked the enemy at Franklin, and drove him from his outer line of temporary works into his interior line which he abandoned during the night, leaving his dead and wounded in our possession, and rapidly retreated to Nashville, closely pursued by our cavalry. We captured several stands of colors and about one thousand (1000) prisoners. Our troops fought with great gallantry. We have to lament the loss of many gallant officers and brave men. Major General Cleburne, Brigadier Generals Gist, John Adams, Strahl, and Granberry, were killed; Major General Brown, Brigadier Generals Carter, Manigault, Quarles, Cockrell, and Scott, were wounded, and Brigadier General Gordon, captured.

"J. B. HOOD, *General.*"

I rode over the scene of action the next morning, and could but indulge in sad and painful thought, as I beheld so many

---

* In an address delivered at Charleston, S. C., I estimated our strength, at Franklin, at twenty-eight thousand (28,000), having overlooked the fact that two of Lee's Divisions could not become engaged.

battle of the 20th. He knew also in what manner my orders at Spring Hill had been totally disregarded. After our last brief interview which was followed so quickly by his death, I sought to account for his sudden revolution of feeling and his hopefulness, since he had been regarded as not over sanguine of the final triumph of our cause. I formed the conviction that he became satisfied on the morning of the 30th of November, after having reviewed the occurrences of the previous afternoon and night, and those of the 20th and 22d of July, that I was not the reckless, indiscreet commander the Johnston-Wigfall party represented me; that I had been harshly judged, and feebly sustained by the officers and men; that I was dealing blows and making moves which had at least the promise of happy results, and that we should have achieved decided success on two occasions around Atlanta as well as at Spring Hill. He therefore made a sudden and firm resolution to support me in all my operations, believing that my movements and manner of handling troops were based upon correct principles. It has been said he stated, upon the morning after the affair of Spring Hill, that he would never again allow one of my orders for battle to be disobeyed, if he could prevent it. For these reasons his loss became doubly great to me. The heroic career and death of this distinguished soldier must ever endear the memory of his last words to his commander, and should entitle his name to be inscribed in immortal characters in the annals of our history.

A similar revolution in feeling took place to a great extent among both officers and men, the morning of the day upon which was fought the battle of Franklin; this change—and in a measure the improved *morale* of the Army, which had resulted from a forward movement of one hundred and eighty miles—occasioned the extraordinary gallantry and desperate fighting witnessed on that field.

The subjoined extract from Van Horne's History of the Army of the Cumberland, will confirm my assertion in regard to our nearly-won victory. Referring to the main breach in the Federal works, the author says: *

"Toward the breach, the enemy's heavy central lines began to press, and to his lateral lines were turned, in seemingly overwhelming convergence. To General Hood, the advantage so easily gained, premised the capture or destruction of the National Army, and he and his Army were inspired to quickest action to maintain and utilize it for this grand achievement. And he certainly could have maintained his hold of the National line, and used for extreme success, had time been given him to thrust into the breach his rapidly advancing and massive rear lines; and as it was, he began to gain ground, right and left, from the Columbia road."

As shown by Colonel Mason's official report, made on the 10th of December, ten days after the battle, our effective strength was: Infantry, eighteen thousand three hundred and forty-two (18,342); artillery, two thousand four hundred and five (2405); cavalry, two thousand three hundred and six (2306); total, twenty-three thousand and fifty-three (23,053). This last number, subtracted from thirty thousand six hundred (30,600), the strength of the Army at Florence, shows a total loss from all causes of seven thousand five hundred and forty-seven (7547), from the 6th of November to the 10th of December, which period includes the engagements at Columbia, Franklin, and of Forrest's cavalry.

The enemy's estimate of our losses as well as of the number of Confederate colors captured is erroneous, as will be seen by the following telegram:

[No. 560.]

"HEADQUARTERS NEAR NASHVILLE, ON FRANKLIN PIKE,
" *December 15th, 1864.*

"HONORABLE J. A. SEDDON, *Secretary of War, Richmond.*

"The enemy claim that we lost thirty colors in the fight at Franklin. We lost thirteen, capturing nearly the same number. The men who bore ours were killed on or within the enemy's interior line of works.

"J. B. HOOD, *General.*"

The estimate of the actual loss at Franklin, given in my

* Van Horne's History, vol. II, pages 199, 200.

official report, was made with the assistance of General Shoupe, my chief of staff, and is, I consider, correct. However, I will estimate later the total loss from all causes, in order to avoid possible error.

After the failure of my cherished plan to crush Schofield's Army before it reached its strongly fortified position around Nashville, I remained with an effective force of only twenty-three thousand and fifty-three. I was therefore well aware of our inability to attack the Federals in their new stronghold with any hope of success, although Schofield's troops had abandoned the field at Franklin, leaving their dead and wounded in our possession, and had hastened with considerable alarm into their fortifications—which latter information, in regard to their condition after the battle, I obtained through spies. I knew equally well that in the absence of the prestige of complete victory, I could not venture with my small force to cross the Cumberland river into Kentucky, without first receiving reinforcements from the Trans-Mississippi Department. I felt convinced that the Tennesseans and Kentuckians would not join our forces, since we had failed in the first instance to defeat the Federal Army and capture Nashville. The President was still urgent in his instructions relative to the transferrence of troops to the Army of Tennessee from Texas, and I daily hoped to receive the glad tidings of their safe passage across the Mississippi river.

Thus, unless strengthened by these long-looked for reinforcements, the only remaining chance of success in the campaign, at this juncture, was to take position, entrench around Nashville, and await Thomas's attack which, if handsomely repulsed, might afford us an opportunity to follow up our advantage on the spot, and enter the city on the heels of the enemy.

I could not afford to turn southward, unless for the *special* purpose of forming a junction with the expected reinforcements from Texas, and with the avowed intention to march back again upon Nashville. In truth, our Army was in that condition which rendered it more judicious the men should face a decisive issue rather than retreat—in other words, rather than renounce the honor of their cause, without having made a last and manful effort to lift up the sinking fortunes of the Confederacy.

I therefore determined to move upon Nashville, to entrench, to accept the chances of reinforcements from Texas, and, even at the risk of an attack in the meantime by overwhelming numbers, to adopt the only feasible means of defeating the enemy with my reduced numbers, viz., to await his attack, and, if favored by success, to follow him into his works. I was apprised of each accession to Thomas's Army, but was still unwilling to abandon the ground as long as I saw a shadow of probability of assistance from the Trans-Mississippi Department, or of victory in battle; and, as I have just remarked, the troops would, I believed, return better satisfied even after defeat if, in grasping at the last straw, they felt that a brave and vigorous effort had been made to save the country from disaster. Such, at the time, was my opinion, which I have since had no reason to alter.

In accordance with these convictions, I ordered the Army to move forward on the 1st of December in the direction of Nashville; Lee's Corps marched in advance, followed by Stewart's and Cheatham's Corps, and the troops bivouacked that night in the vicinity of Brentwood. On the morning of the 2d, the march was resumed, and line of battle formed in front of Nashville. Lee's Corps was placed in the centre and across the Franklin pike; Stewart occupied the left, and Cheatham the right—their flanks extending as near the Cumberland as possible, whilst Forrest's cavalry filled the gap between them and the river.

General Rousseau occupied Murfreesboro', in rear of our right, with about eight thousand men heavily entrenched. General Bates's Division, Sears's and Brown's brigades, were ordered, on the 5th, to report at that point to General Forrest, who was instructed to watch closely that detachment of the enemy. The same day, information was received of the capture of one hundred prisoners, two pieces of artillery, twenty wagons and teams by Forrest's cavalry, at Lavergne; of the capture and destruction of three block houses on the Chattanooga Railroad, by Bates's Division; and of the seizure the day previous, by General Chalmers, of two transports on the Cumberland river, with three hundred mules on board.

We had in our possession two engines and several cars, which ran as far south as Pulaski. Dispatches were sent to Generals Beauregard and Maury to repair the railroad from Corinth to Decatur, as our trains would be running in a day or two to the latter point. This means of transportation was of great service in furnishing supplies to the Army. Our troops had, when we reached Middle Tennessee, an abundance of provisions, although sorely in need of shoes and clothing.

At this time, I telegraphed the War Department to request that General Breckinridge's command, in West Virginia, be sent to me or ordered into Kentucky to create a diversion and lessen the concentration of the Federal Army in my front. General R. E. Lee's necessities were, however, more urgent than my own. The application was, therefore, not granted.

On the 7th, intelligence was received, and telegraphed to General Beauregard, that General Steele, with fifteen thousand (15,000) troops, had passed Memphis in the direction of Cairo; also, that Rousseau had made a sally, and driven back our forces at Murfreesboro'. The following day General Forrest was instructed to leave the roads open to Lebanon, in the hope of enticing Rousseau out of his stronghold; preparations were at the same time made to capture his detachment of eight thousand, should he venture to reinforce Thomas at Nashville. He remained, however, behind his entrenchments.

General Bates's Division was ordered to return to the Army; Forrest was instructed to direct Palmer's and Mercer's infantry brigades to thoroughly entrench on Stewart's creek, or at Lavergne, according as he might deem more judicious;

to constitute, with these troops and his cavalry, a force in observation of the enemy at Murfreesboro', and, lastly, to send a brigade of cavalry to picket the river at Lebanon.

The Federals having been reported to be massing cavalry at Edgefield, Forrest was instructed to meet and drive them back, if they attempted to cross the Cumberland. The same day, the 10th of December, Generals Stewart and Cheatham were directed to construct detached works in rear of their flanks, which rested near the river, in order to protect these flanks against an effort by the Federals to turn them. Although every possible exertion was made by these officers, the works were not completed when, on the 15th, the Federal Army moved out, and attacked both flanks, whilst the main assault was directed against our left. It was my intention to have made these defences self-sustaining, but time was not allowed, as the enemy attacked on the morning of the 15th. Throughout that day, they were repulsed at all points of the general line with heavy loss, and only succeeded towards evening in capturing the infantry outposts on our left, and with them the small force together with the artillery posted in these unfinished works.

Finding that the main movement of the Federals was directed against our left, the chief engineer was instructed to carefully select a line in prolongation of the left flank; Cheatham's Corps was withdrawn from the right during the night of the 15th, and posted on the left of Stewart—Cheatham's left flank resting near the Brentwood Hills. In this position, the men were ordered to construct breastworks during that same night.

The morning of the 16th found us with Lee's right on Overton Hill. At an early hour the enemy made a general attack along our front, and were again and again repulsed at all points with heavy loss, especially in Lee's front. About 3.30 p. m. the Federals concentrated a number of guns against a portion of our line, which passed over a mound on the left of our centre, and which had been occupied during the night.

This point was favorable for massing troops for an assault under cover of artillery. Accordingly the enemy availed himself of the advantage presented, massed a body of men—apparently one division—at the base of this mound, and, under the fire of artillery, which prevented our men from raising their heads above the breastworks, made a sudden and gallant charge up to and over our entrenchments. Our line, thus pierced, gave way; soon thereafter it broke at all points, and I beheld for the first and only time a Confederate Army abandon the field in confusion.

Major General Bates, in his official report, refers to an angle having been formed upon the mound where the line first gave way. If such be the case, the officers in command of the troops at that point were doubtless at fault, as Colonel Prestman, chief engineer, and his assistants, had staked off the line with great care, and I am confident were not guilty of this *grave neglect.* I was seated upon my horse not far in rear when the breach was effected, and soon discovered that all hope to rally the troops was vain.

I did not, I might say, anticipate a break at that time, as our forces up to that moment had repulsed the Federals at every point, and were waving their colors in defiance, crying out to the enemy, "Come on, come on." Just previous to this fatal occurrence, I had matured the movement for the next morning. The enemy's right flank, by this hour, stood in air some six miles from Nashville, and I had determined to withdraw my entire force during the night, and attack this exposed flank in rear. I could safely have done so, as I still had open a line of retreat.

The day before the rout, the artillery posted in the detached works had been captured; a number of guns in the main line were abandoned at the time of the disaster, for the reason that the horses could not be brought forward in time to remove them. Thus the total number of guns captured amounted to fifty-four.

We had fortunately still remaining a sufficient number of

pieces of artillery for the equipment of the Army, since, it will be remembered, I had taken with me at the outset of the campaign a large reserve of artillery to use against gunboats. Our losses in killed and wounded in this engagement were comparatively small, as the troops were protected by breastworks.

An incident at the time of the rout was reported to me which I deem worthy of mention. When our troops were in the greatest confusion, a young lady of Tennessee, Miss Mary Bradford, rushed in their midst regardless of the storm of bullets, and, in the name of God and of our country, implored them to re-form and face the enemy. Her name deserves to be enrolled among the heroes of the war, and it is with pride that I bear testimony to her bravery and patriotism.

Order among the troops was in a measure restored at Brentwood, a few miles in rear of the scene of disaster, through the promptness and gallantry of Clayton's Division, which speedily formed and confronted the enemy, with Gibson's brigade and McKenzie's battery, of Fenner's battalion, acting as rear guard of the rear guard. General Clayton displayed admirable coolness and courage that afternoon and the next morning in the discharge of his duties. General Gibson, who evinced conspicuous gallantry and ability in the handling of his troops, succeeded, in concert with Clayton, in checking and staying the first and most dangerous shock which always follows immediately after a rout. The result was that even after the Army passed the Big Harpeth, at Franklin, the brigades and divisions were marching in regular order. Captain Cooper, of my staff, had been sent to Murfreesboro' to inform General Forrest of our misfortune, and to order him to make the necessary dispositions of his cavalry to cover our retreat.

Although the campaign proved disastrous by reason of the unfortunate affair at Spring Hill, the short duration of daylight at Franklin, and, finally, because of the non-arrival of the expected reinforcements from the Trans-Mississippi Department, it will nevertheless be of interest to note how deeply

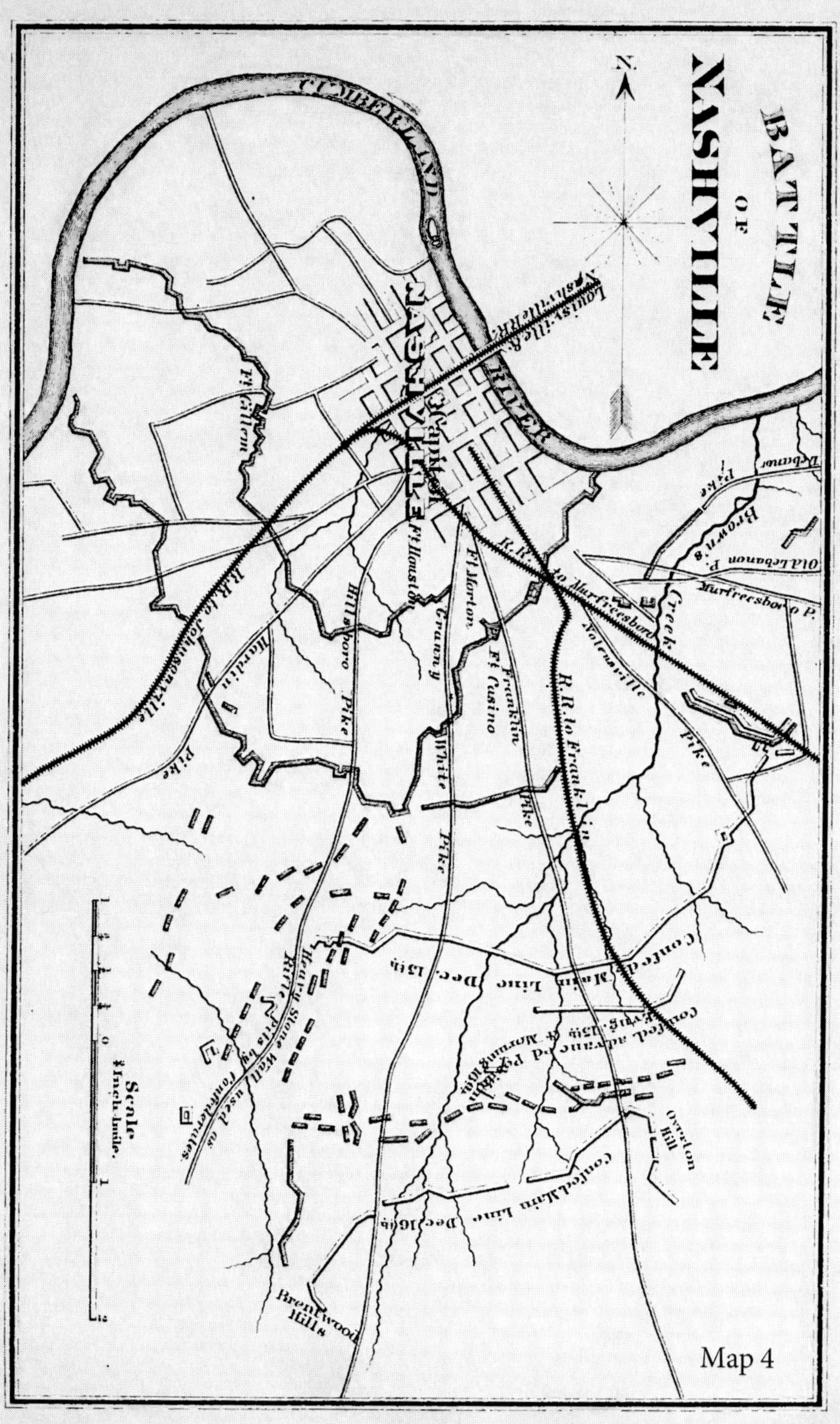

BATTLE OF NASHVILLE
N.
CUMBERLAND RIVER
NASHVILLE
Capitol
Louisville R.R.
Nashville R.R.
Ft. Gillem
Ft. Houston
Ft. Morton
Ft. Casino
R.R. to Johnsonville
Harding Pike
Hillsboro Pike
Granny White Pike
Franklin Pike
R.R. to Franklin
R.R. to Murfreesboro
Nolensville Pike
Murfreesboro P.
Old Lebanon P.
Lebanon Pike
Brown's Creek
Confed. Main Line Dec. 15th
Confed. Main Line Dec. 16th
Heavy Stone Wall used as Rifle pits by Confederates
Overton Hill
Brentwood Hills
Scale
1 Inch-1 mile.

Map 4

concerned General Grant became for fear we should finally reach Kentucky. He ordered General Thomas to attack on the 6th of December, and evidently became much worried about our presence in front of Nashville, as he telegraphed to the War Department at Washington, on the 9th, to relieve Thomas on account of his delay in assaulting according to instructions. This order was issued on that date, but was afterwards suspended by Grant. On the 11th, at 4 p. m., he again telegraphed General Thomas.*

"If you delay attacking longer, the mortifying spectacle will be witnessed of a rebel Army moving for the Ohio, and you will be forced to act, accepting such weather as you find. * * * *"

The following dispatch from General Grant to Thomas gives strong evidence that in this campaign we had thrust at the vitals of the enemy: †

"WASHINGTON, *December 15th, 1864, 11.30 p. m.*

"I was just on my way to Nashville, but receiving a dispatch from Van Duzen, detailing your splendid success of to-day, I shall go no further. * * *

"U. S. GRANT, *Lieutenant General.*"

He could not well afford to allow us to reach Kentucky, and finally assail him in rear at Petersburg. Therefore he left his own Army in front of the illustrious Lee to proceed to Nashville and assume direction in person.

At this eventful period General Thomas stood with eighty-two thousand (82,000) effectives ‡ to oppose our small Army, which numbered less than twenty thousand (20,000) after deducting the force under Forrest at Murfreesboro'.

I had had reason to hope that we would have received large accessions to our ranks in Tennessee. The following letter from Governor Isham G. Harris, written during the retreat and at the time the Army was approaching the Tennessee river,

* Van Horne's History Army of the Cumberland, vol. II, page 257.

† Van Horne's History, vol. II, page 259.

‡ Sherman's Memoirs, vol. II, pages 162, 163.

will indicate to what extent our ranks would have been recruited, had the campaign proved successful:

"TUSCUMBIA, ALABAMA, *December 25th, 1864.*

"HIS EXCELLENCY, JEFFERSON DAVIS.

"SIR:—I arrived here last night, leaving the Army some fifteen miles beyond the Tennessee river, on the Bainbridge route.

"Our stay in Tennessee was so short, and engagements so constant and pressing that we did not recruit to any considerable extent. If we could have remained there a few weeks longer, we could and would have recruited to a great extent. The men are there, and thousands were making their arrangements to join the Army, but the unfortunate result of the battle of Nashville, and immediate retreat of the Army was very discouraging to our people. I hope, however, to be able to get a great many of these men out, notwithstanding we have left the State.

"I have been with General Hood from the beginning of this campaign, and beg to say, disastrous as it has ended, I am not able to see anything that General Hood has done that he should not, or neglected anything that he should have done which it was possible to do. Indeed, the more that I have seen and known of him and his policy, the more I have been pleased with him and regret to say that if *all* had performed their parts as well as he, the results would have been very different.

"But I will not detain Colonel Johnson, except to say or rather to suggest that if General Hood is to command this Army, he should by all means be permitted to organize the Army according to his own views of the necessities of the case.

"Very respectfully,
"ISHAM G. HARRIS."

Lieutenant General Lee displayed his usual energy and skill in handling his troops on the 17th, whilst protecting the rear of our Army. Unfortunately, in the afternoon he was wounded and forced to leave the field. Major General Carter L. Stevenson then assumed command of Lee's Corps, and ably discharged his duties during the continuance of the retreat to and across the Tennessee river.

Major General Walthall, one of the most able division commanders in the South, was here ordered to form a rear guard with eight picked brigades together with Forrest's cavalry; the march was then resumed in the direction of Columbia, Stewart's Corps moving in front, followed by those of Cheatham and

Stevenson. The Army bivouacked in line of battle near Duck river, on the night of the 18th.

The following day, we crossed the river and proceeded on different roads leading towards Bainbridge on the Tennessee. I entertained but little concern in regard to being further harassed by the enemy. I felt confident that Walthall, supported on his flanks by the gallánt Forrest, would prove equal to any emergency which might arise. I therefore continued, although within sound of the guns of the rear guard, to march leisurely, and arrived at Bainbridge, on the 25th of December. A pontoon bridge was constructed as rapidly as the boats arrived, the corps were placed in position covering the roads to the north, and during the 26th and 27th the Army crossed the river. The following day, the march was continued in the direction of Tupelo, at which place Cheatham's Corps, the last in the line of march, went into camp on the 10th of January, 1865.

I had telegraphed General Beauregard from Bainbridge to meet me, and, in compliance with my request, he arrived at Army headquarters on the night of the 14th. The day previous, I had sent the following dispatch to the Secretary of War:

"HEADQUARTERS, TUPELO, MISSISSIPPI,
"*January 13th, 1865.*

"HONORABLE J. A. SEDDON, *Secretary of War, Richmond.*

"I request to be relieved from the command of this Army.

"J. B. HOOD, *General.*"

On the 15th, after consultation with General Beauregard, a system of furloughing the troops was agreed upon. In reference thereto, I find the following memorandum in General Shoupe's diary:

"A system of furloughing the troops established. See General Order No. 1, 1865, and circular letter to corps commanders, field dispatches, No. 542."

In a dispatch of January 3d to President Davis, I asked for authority to grant a leave of absence to the Trans-Mississippi troops; and, as the men from Tennessee had stood by their colors notwithstanding the Army had been forced to abandon their State, I deemed it wise, in consideration of their faithful services, to at least grant them a short leave of absence, as well as to others who might be able to go home and return within ten or fifteen days. General Beauregard concurred with me, and the general order above referred to was issued, as the ensuing circular will indicate:

[No. 542.]

"HEADQUARTERS, TUPELO, MISSISSIPPI,
"*January 16th, 1865.*

"LIEUTENANT GENERAL STEWART, MAJOR GENERAL STEVENSON, MAJOR GENERAL CHEATHAM.

(Copy sent to Colonel Harvie.)

"If you have any troops in your command who live sufficiently near the present position of the Army to justify, in your judgment, the granting them ten days' furlough, the same will be done on proper application made at once, provided the men go by organizations under officers, and pledge themselves to return at the expiration of the time. All obtaining such furloughs will be debarred the benefit of General Order No. 1 from Army headquarters.

"By command of General Hood,

"A. P. MASON,

"Lieutenant Colonel, A. A. G."

I regret that I have not this general order in my possession. My recollection is quite clear, however, that it referred in a great measure to the furloughing of the Tennessee troops—about two thousand in number—and of those who lived in the vicinity. It is a source of equal regret to me that I have not the field return of the Army, which was being made upon the 23d of January, the day I left Tupelo for Richmond. The following letter from Colonel A. P. Mason, assistant adjutant general, written soon thereafter, will establish the approximate strength of the Army after its arrival at Tupelo, on the 10th of January:

[Private.]

"RICHMOND, *March 10th, 1865.*

"GENERAL:—In compliance with your request made a few days since in reference to the strength of the Army of Tennessee at the time you left

Tupelo, Mississippi, I respectfully submit that, according to my recollection of a 'Field Return' of the Army, which was being made at that time and finished a day or two after your departure, the 'effective total' of the infantry and artillery was about fifteen thousand (15,000)—perhaps a few hundred less. This return was made after the West Tennessee regiments of Major General Cheatham's Corps had been furloughed, as well as some men furloughed under an order published at Tupelo, and some small organizations also furloughed at Tupelo. I cannot form any estimate of the number of men thus furloughed, because you will remember that all the organization furloughs were given by the corps commanders (your sanction having been previously obtained); consequently the strength of such organizations, at the time they were furloughed, was not furnished the assistant adjutant general's office at Army headquarters.

"The 'Field Return' above referred to was sent to Colonel Brent, and was in his office in Augusta when I passed there a few weeks since.

"Most respectfully your obedient servant,

"A. P. MASON,

"Lieutenant Colonel, A. A. G."

Under the foregoing order not less than three thousand five hundred (3500) men were furloughed prior to the date upon which the return was made up. Now since Colonel Mason was the adjutant general under whose direction it was made, there can hardly be any question but that the Army, after its arrival at Tupelo, numbered from eighteen thousand (18,000) to nineteen thousand (19,000) effective troops of the infantry and artillery. General D. H. Maury, commanding at that period in Mobile, informs me by letter that about four thousand (4000) of these forces joined him from Tupelo, armed and equipped. General Johnston states in his Narrative that only about five thousand (5000) reached him in North Carolina, and, adducing the oral statement of two officers, endeavors to create the impression that their arms had been lost, and that this remnant constituted the Army of Tennessee at the time I relinquished its command. Whereas—notwithstanding the outcry against me, and the general declaration through the press that, if Johnston were restored to command, absentees and deserters would return by the thousand and our independence be secured, and although it was understood, before my departure from Tupelo, that he would be reinstated —nine thousand out of fourteen thousand, who left Tupelo to repair to his standard in North Carolina, deserted, and either went to the woods or to their homes. This affords positive proof that General Beauregard and I judged aright at Gadsden and also at Florence, Alabama, in regard to the Army, when we decided that to turn and follow Sherman would cause such numbers to desert, as to render those who were too proud to quit their colors almost useless.

In accordance with Colonel Mason's letter of March the 10th, there were, including the furloughed men, about eighteen thousand five hundred (18,500) effectives of the infantry and artillery at Tupelo, after my retreat from Nashville; and it will be seen in his return of November 6th, which date was near the time of our advance into Tennessee, that the effective strength of the Army at that period was thirty thousand six hundred (30,600), inclusive of the cavalry.

Thus we find at Tupelo eighteen thousand five hundred (18,500) infantry and artillery, and twenty-three hundred and six (2306) of Forrest's cavalry; to which add ten thousand lost from all causes, and the sum total amounts to thirty thousand eight hundred and six (30,806) effectives, which proves my loss during the Tennessee campaign to have been not in excess of ten thousand (10,000), as I announced in my official report. As previously mentioned, Wheeler's cavalry, reported at ten thousand (10,000), was left in Georgia when I marched into Tennessee, and was replaced by Forrest's cavalry, which accompanied the Army.

Upon General Beauregard's arrival at Tupelo, on the 14th of January, I informed him of my application to be relieved from the command of the Army. As the opposition of our people, excited by the Johnston-Wigfall party, seemingly increased in bitterness, I felt that my services could no longer be of benefit to that Army; having no other aspiration than to promote the interests of my country, I again telegraphed the authorities in Richmond, stating that the campaigns to the Alabama line and into Tennessee were my own conception;

that I alone was responsible; that I had striven hard to execute them in such manner as to bring victory to our people, and, at the same time, repeated my desire to be relieved. The President finally complied with my request, and I bid farewell to the Army of Tennessee on the 23d of January, 1865, after having served with it somewhat in excess of eleven months, and having performed my duties to the utmost of my ability.

At the time I assumed command around Atlanta, a number of General Johnston's staff officers remained with me, among whom were Colonels Mason, Falconer and Harvie, Majors Henry and Clare, who, notwithstanding the extraordinary circumstances under which I had superseded their old commanding officer, ably discharged their various duties with zeal and strict fidelity.

After leaving Tupelo, I returned to Virginia and found President Davis still most anxious to procure reinforcements from the Trans-Mississippi Department. He consulted fully with General Lee in regard to this important matter, and, after a sojourn of several weeks in Richmond, during which interval I prepared my official report, I was ordered to Texas with instructions to gather together all the troops willing to follow me from that State, and move at once to the support of General Lee. Soon after my arrival at Sumpter, South Carolina, I received the painful intelligence of Lee's surrender. Nevertheless, I continued my journey, and about the last of April reached the Mississippi, in the vicinity of Natchez. Here I remained with my staff and escort, using vain endeavors to cross this mighty river, until after the receipt of positive information of General E. Kirby Smith's surrender. During this interim we were several times hotly chased by Federal cavalry through the wood and canebrake. Finally, on the 31st of May, 1865, I rode into Natchez and proffered my sword to Major General Davidson, of the United States Army. He courteously bade me retain it, paroled the officers and men in company with me, and allowed us to proceed without delay to Texas, via New Orleans.

CHAPTER XVIII.

RASHNESS—JOHNSTON—FABIUS—SCIPIO.

Before closing these pages, I request the privilege of correcting a false impression which has gained ground in my regard, and which is, I may say, the outcome of inimical statements of certain writers who have followed in the wake of Pollard and Johnston.

General Sherman gives color to their charge of rashness as a commander, in the following passage:

> "I did not suppose that General Hood, though rash, would venture to attack fortified places like Allatoona, Resaca, Decatur and Nashville; but he did so, and in so doing, played into our hands perfectly." *

And yet from other portions of his Memoirs it will be seen that I did not attack either Resaca, Decatur, or Nashville. My official report will also show that Major General French assaulted Allatoona, whilst under discretionary orders. Thus, in none of these instances is General Sherman correct.

Touching this same accusation of rashness, put forth by my opponents, I shall merely state that the confidence reposed in me upon so many occasions, and during a service of three years, by Generals Lee, Jackson, and Longstreet, in addition to the letters of these distinguished commanders, expressive

* Sherman's Memoirs, vol. II, page 167.

of satisfaction with my course, is a sufficient refutation of the charge.

The above allegation is not more erroneous than the following inference is illogical. Van Horne, in his History of the Army of the Cumberland, speaks in commendation of my movement to the rear of Sherman, after the fall of Atlanta, but regards the circumstance as unfortunate for the Confederacy that Johnston was not summoned to Palmetto at the beginning of the new campaign, in order to insure its successful issue. The writer must assuredly have been ignorant of the antecedents of this General when he formed this conclusion; it seems, indeed, preposterous to suppose that General Johnston would have inaugurated a similar movement with thirty-five thousand (35,000) men, when he had just retreated from the same territory with an Army of seventy thousand (70,000) and when he had declined to make, with an effective force of over eighty thousand (80,000), the same campaign from Dalton the preceding Spring.

Now, since I have been charged with rashness, and even recklessness, by General Johnston and his adherents, I may be allowed, in addition to answering this severe arraignment, to at least question his right to be considered one of our leading Generals.

It has been asserted that he pursued the Fabian policy in his campaign from Dalton to Atlanta. It is, indeed, to be regretted that he did not follow in the footsteps of the renowned Roman by holding on to the mountains of Georgia. In the long course of years, during which Fabius Maximus commanded at intervals the Roman Legions, he could never be induced to quit the mountainous regions, and accept the gage of battle with Hannibal upon the plains. Neither the taunts nor stratagems of his enemy, nor the contempt and ridicule of his own people, could make him depart from his resolution, and abandon the heights. The people finally grew so dissatisfied under his policy that he was required to share the command of the Army with Minucius. During the long and eventful period embraced in the second Punic war, which lasted eighteen years, different commanders sallied forth and delivered battle; but Fabius continued to adhere strictly to his plan of warfare, and stubbornly refused to encounter his antagonist in the plains. His colleague, Minucius, an imprudent and even rash General, dashed down from the heights with one-half of the Army, engaged Hannibal, and was only spared utter destruction by the timely aid of Fabius. Varro marched out, fought the Carthagenians near Cannæ, was defeated, and left forty thousand Romans upon the field. Marcellus, a more fortunate General, gained important advantages over the enemy; but, as history tells us, Fabius permitted no allurement of his foe, nor outcry of his countrymen, to induce him to descend from the mountains.

His policy was, seemingly, as fixed and unchangeable as the sun in the eternal heavens. Plutarch relates that in order "to secure himself against the enemy's horse, he took care to encamp above them on high and mountainous places. When they sat still, he did the same; when they were in motion, he showed himself upon the heights, at such a distance as not to be obliged to fight against his inclination, and yet near enough to keep them in perpetual alarm, as if, amidst his arts to gain time, he intended every moment to give them battle. These dilatory proceedings exposed him to contempt among the Romans in general, and even in his own Army. * * * Thus the soldiers were brought to despise Fabius, and by way of derision to call him the *pedagogue* of Hannibal, while they extolled Minucius as a great man and one that acted up to the dignity of Rome. This led Minucius to give a freer scope to his arrogance and pride, and to ridicule the Dictator for encamping constantly upon the mountains 'as if he did it on purpose that his men might more clearly behold Italy laid waste with fire and sword.' And he asked the friends of Fabius 'whether he intended to take his Army up into heaven, as if he had bid adieu to the world below, or whether he would screen himself from the enemy with clouds and fogs?'

When the Dictator's friends brought him an account of these aspersions, and exhorted him to wipe them off by risking a battle: 'In that case,' said he, 'I should be of a more dastardly spirit than they represent me, if through fear of insults and reproaches, I should depart from my own resolution.'"

Therefore when General Johnston retreated from the mountain-fastnesses, crossed the Chattahoochee river, and moved out upon the plains of Georgia, he bid adieu forever to even a shadow of right to the claim of having pursued the policy so persistently carried out by Fabius Maximus. Had he clung to the mountains and refused to surrender them to General Sherman, vast indeed might have been the results achieved, and far greater his title to distinction. Although Fabius succeeded in wasting in a great measure the strength of his adversary, it however required the boldness and the genius of Scipio to finally defeat Hannibal, and place Carthage beneath the heel of the proud Roman.

General Johnston not only signally failed in the Fabian policy, but, unfortunately, declined to act the part of Scipio Africanus, at Dalton, in the early Spring of 1864.

History records the deeds of this famed warrior who, whilst the Carthagenians were still warring in Italy, aroused the Roman pride, gathered together his legions, moved to the rear of the enemy, transferred the war into Africa, forced the recall of Hannibal, routed his Army in battle, placed Carthage at his feet, and brought security and prosperity to his countrymen. Arnold, in his History of Rome, gives a lengthy and interesting description of this bold and brilliant move, and of the victories which followed. Plutarch condenses the whole into these few words: "After Scipio was gone over into Africa, an account was soon brought to Rome of his glorious and wonderful achievements. This account was followed by rich spoils which confirmed it. A Numidian king was taken prisoner; two camps were burned and destroyed, and in them a vast number of men, arms and horses; and the Carthagenians sent orders to Hannibal to quit his fruitless hopes in Italy, and return home to defend his own country. * * Soon after, Scipio defeated Hannibal in a pitched battle, pulled down the pride of Carthage, and trod it under foot. This afforded the Romans a pleasure beyond all their hopes, and restored a firmness to their empire, which had been shaken by so many tempests."

Since General Johnston failed to practice the art of war in accordance with the principles either of Fabius Maximus or of Scipio Africanus, and since he fought not a single *general* battle during the entire war of Secession, what just claim has he to generalship? A man may be learned in anatomy, and perfect in the theory of surgery; he may be able to thoroughly describe the manner in which the most difficult operation should be performed, but may never have possessed sufficient nerve to undertake even one operation in which serious risk was involved, and thus give life to his theories by practical work. Who would employ a surgeon who had never used the knife? Furthermore, who could, under the circumstances, declare him with reason an eminent man in his profession? Ruskin can, probably, better describe a painting than any artist of ancient or modern times. His gorgeous descriptions attracted the attention of the world to the wonderful genius of Turner; but who would venture to assert that he himself was a great painter, when he has perhaps never used the brush? Thus it is as it should be: no man is justly entitled to be considered a great General, unless he has won his spurs. Had General Johnston possessed the requisite spirit and boldness to seize the various chances for victory, which were offered him, he never would have allowed General Sherman to push him back one hundred miles in sixty-six days, from one mountain stronghold to another, down into the very heart of the Confederacy.

APPENDIX.

GENERAL HOOD'S REPORT.

THE OPERATIONS OF THE ARMY OF TENNESSEE.

RICHMOND, VA., *Feb. 15th, 1865.*

GENERAL S. COOPER,

Adjutant and Inspector General, Richmond, Va.

GENERAL:—I have the honor to submit the following Report of the operations of the Army of Tennessee, while commanded by me, from July 18th, 1864, to January 23d, 1865.

The results of a campaign do not always show how the General in command has discharged his duty. Their enquiry should be not what he has done, but what he should have accomplished with the means under his control. To appreciate the operations of the Army of Tennessee, it is necessary to look at its history during the three months which preceded the day on which I was ordered to its command. To do this, it is necessary either to state in this report all the facts which illustrate the entire operations of the Army of Tennessee in the recent campaign, or to write a supplemental or accompanying report. I deem the former more appropriate, and will therefore submit in a single paper all the information which seems to me should be communicated to the Government.

On the 6th of May, 1864, the Army lay at and near Dalton awaiting the advance of the enemy.

Never had so large a Confederate Army assembled in the West. Seventy thousand (70,000) effective men were in the easy direction of a single commander, whose good fortune it was to be able to give successful battle, and redeem the losses of the past. Extraordinary efforts had been used to secure easy victory. The South had been denuded of troops to fill the strength of the Army of Tennessee. Mississippi and Alabama were without military support, and looked for protection in decisive battle in the mountains of Georgia. The vast forces of the enemy were accumulating in the East, and to retard their advance or confuse their plans, much was expected by a counter movement by us in the West. The desires of the Government expressed to the Confederate commander in the West were to assume the offensive. Nearly all the men and resources of the West and South were placed at his disposal for the purpose. The men amounted to the number already stated, and the resources for their support were equal to the demand. The reinforcements were within supporting distance. The troops felt strong in their increased numbers, saw the means and arrangements to move forward, and recover, not abandon our own territory, and believed that victory might be achieved. In such condition was that splendid Army when the active campaign fairly opened. The enemy but little superior in numbers, none in organization and discipline, inferior in spirit and confidence, commenced his advance. The Confederate forces, whose faces and hopes were to the North, almost simultaneously commenced to retreat. They soon reached positions favorable for resistance. Great ranges of mountains running across the line of march, and deep rivers, are stands from which a well-directed Army is not easily driven or turned. At each advance of the enemy the Confederate Army, without serious resistance, fell back to the next range or river in the rear. This habit to retreat soon became a routine of the Army, and was substituted for the hope and confidence with which the campaign opened. The enemy soon perceived this. With perfect security he divided his forces, using one column to menace in front, and one to threaten in rear. The usual order to retreat, not strike in detail, was issued and obeyed. These retreats were always at night,—the day was consumed in hard labor. Daily temporary works were thrown up, behind which it was never intended to fight. The men became travellers by night, and laborers by day. They were ceasing to be soldiers by the disuse of military duty. Thus for seventy-four days and nights that noble Army, if ordered to resist, no force that the enemy could assemble could dislodge from a battle field, continued to abandon their country, to see their strength departing, and their flag waving only in retreat, or in partial engagements. At the end of that time after descending from the mountains when the last advantage of position was abandoned, and camping, without fortifications, on the open plains of Georgia, the Army had lost twenty-two thousand seven hundred and fifty (22,750) of its best soldiers. Nearly one-third was gone, no general battle fought, much of our State abandoned, two others uncovered, and the organization and efficiency of every command by loss of officers, men, and tone, seriously diminished. These things were the inevitable result of the strategy adopted. It is impossible for a large Army to retreat in the face of a pursuing enemy without such a fate. In a retreat the losses are constant and permanent. Stragglers are overtaken; the fatigued fall by the wayside, and are gathered by the advancing enemy. Every position by the rear guard, if taken, yields its wounded to the victors. The soldiers always awaken from rest at night to continue the retreat, leave many of their comrades asleep in trenches. The losses of a single day are not large. Those of seventy-four days will embrace the strength of an Army. If a battle be fought, and the field held at the close, however great the slaughter, the loss will be less than to retreat in the face of an enemy. There will be no stragglers. Desertions are in retreat, rarely, if ever, on the field of battle. The wounded are gathered to the rear, and soon recover, and in a few weeks the entire loss consists only of the killed and permanently disabled, which is not one-fifth of the apparent loss on the night of the battle. The enemy is checked, his plans deranged, territory saved, the campaign suspended or won. If a retreat still be necessary, it can then be done with no enemy pressing, and no loss following. The advancing party loses nothing but its killed and permanently disabled. Neither straggler nor deserter thin its ranks. It reaches the end of its march stronger for battle than when it started. The Army commanded by General Sherman, and that commanded by General Johnston, not greatly unequal at the commencement of the campaign, illustrate what I have written. General Sherman, in his official report, states that his forces when they entered Atlanta were nearly the same in number as when they left Dalton. The Army of Tennessee lost twenty-two thousand seven hundred and fifty (22,750) men, nearly one-third of its strength; I have nothing to say of the statement of losses made by General Johnston in his official report, except to state that by his own figures he understates his loss some thousands, that he excludes the idea of any prisoners, although his previous official returns show more than seven thousand (7000) under the head, "absent without leave," and that the returns of the Army while he was in command, corrected and increased by the *records* of the Army which has not been fully reported to the Government, and the return signed by me, but made up under him as soon as I assumed command, show the losses of the Army of Tennessee to be what I have stated, and a careful examination of the returns with the Army will show the losses to be more than stated.

This statement of the previous conduct of the campaign is necessary, so as to show what means I had to retrieve the disasters of the past. And if the results are not such as to bring joy to the country, it is not the first time that the most faithful efforts of duty were unable to repair the injury done by others. If, as is untruly charged, the Army of Tennessee ceased to exist under my command, it is also true that it received its mortal wound when it turned its back in retreat in the mountains of Georgia; and under different management it lingered much longer than it would have done with the same daily loss occurring, when it was placed under my direction.

The Army was turned over to me by order of the President at Atlanta on the 18th July, 1864. Its effective strength was: Infantry, thirty-three thousand seven hundred and fifty (33,750); artillery, three thousand five hundred (3500); cavalry, ten thousand (10,000), with one thousand five hundred Georgia militia, commanded by Major General G. W. Smith, making a total effective of forty-eight thousand seven hundred and fifty (48,750) men. The Army was in bivouac south of the Chattahoochee river, between Atlanta and that river, and was advancing—the right near Pace's Ferry and the left near Roswell. On the evening of the 18th our cavalry was principally driven across Peach Tree creek. I caused line of battle to be formed, the left resting near the Pace's Ferry road, and the right covering Atlanta. On the morning of the 19th the dispositions of the enemy were substantially as follows: The "Army of the Cumberland," under Thomas, was in the act of crossing Peach Tree creek. This creek forming a considerable obstacle to the passage of an Army, runs in a northeasterly direction, emptying into the Chattahoochee river near the railroad crossing. The "Army of the Ohio," under Schofield, was also about to cross east of the Buckhead road. The "Army of the Tennessee," under McPherson, was moving on the Georgia Railroad at Decatur. Feeling it impossible to hold Atlanta without giving battle, I determined to strike the enemy while attempting to cross this stream. My troops were disposed as follows: Stewart's Corps on the left; Hardee's in the centre, and Cheatham's on the right, entrenched. My object was to crush Thomas's Army before he could fortify himself, and then turn upon Schofield and McPherson. To do this, Cheatham was ordered to hold his left on the creek in order to separate Thomas's Army from the forces on his (Thomas's) left. Thus I should be able to throw two corps, Stewart's and Hardee's, against Thomas. Specific orders were carefully given these Generals, in the presence of each other, as follows: The attack was to begin at 1 p. m.; the movement to be by division in *en echelon* from the right, at the distance of about one hundred and fifty yards; the effort to be to drive the enemy back to the creek, and then towards the river, into the narrow space formed by the river and creek, everything on our side of the creek to be taken at all hazards, and to follow up as our success might permit. Each of these Generals was to hold a division in reserve. Owing to the demonstrations of the enemy on the right, it became necessary to extend Cheatham a division front to the right. To do this, Hardee and Stewart were each ordered to extend a half division front to close the interval. Foreseeing that some confusion and delay might result, I was careful to call General Hardee's attention to the importance of having a staff officer on his left, to see that the left did not take more than half a division front. This unfortunately was not attended to, and the line closed to the right, causing Stewart to move two or three times the proper distance. In consequence of this, the attack was delayed until nearly 4 p. m. At this hour the attack began as ordered. Stewart's Corps carrying the temporary works in his front; Hardee's failed to push the attack as ordered, and thus the enemy remaining in possession of his works on Stewart's right, compelled Stewart by an enfilade fire to abandon the position he had carried. I have every reason to believe that our attack would have been successful had my order been executed. I am strengthened in this opinion by information since obtained through Brigadier General Govan, some time a prisoner in the enemy's hands, touching the condition of the enemy at the time. The delay from one to four o'clock, p. m., was unfortunate, but would have not proved irretrievable had the attack been vigorously made. Ascertaining that the attack had failed, I caused the troops to retire to their former positions.

The position and demonstration of McPherson's Army on the right threatening my communications, made it necessary to abandon Atlanta or check his movements. Unwilling to abandon, the following instructions were given on the morning of the 21st: The chief engineer was instructed to select a line of defence immediately about Atlanta, the works already constructed for the defence of the place being wholly useless from their position; Stewart's and Cheatham's Corps to take position and construct works to defend the city—the former on the left, the latter on the right. The artillery, under the command of Brigadier General Shoupe, was massed on the extreme right. Hardee was ordered to move with his corps, during the night of the 21st, south on the McDonough road, crossing Entrenchment creek at Cobb's Mills, and to completely turn the left of McPherson's Army. This he was to do even should it be necessary to go to or beyond Decatur. Wheeler, with his cavalry, was ordered to move on Hardee's right, both to attack at daylight, or as soon thereafter as possible. As soon as Hardee succeeded in forcing back the enemy's left, Cheatham was to take up the movement from his right, and continue to force the whole from right to left, down Peach Tree creek, Stewart in like manner to engage the enemy as soon as the movement became general. Hardee failed to entirely turn the enemy's left, as directed, took position, and attacked his flank. His troops fought with great spirit and determination, carrying several lines of entrenchments, Wheeler attacking on the right. Finding Hardee so hotly engaged, and fearing the enemy might concentrate upon him, I ordered Cheatham forward to create a diversion. Hardee held the ground he gained. Cheatham carried the enemy's entrenchments in his front, but had to abandon them in consequence of the enfilade fire brought to bear upon him. Cheatham captured five guns and five or six stands of colors, and Hardee eight guns and thirteen stands of colors. While the grand results desired were not accomplished, the movements of McPherson upon my communications were entirely defeated, and no further effort was made in that direction at any time. This engagement greatly inspired the troops and revived their confidence. Here, I regret to say, the brave and gallant Major General W. H. T. Walker was killed. The enemy withdrew his left to the Georgia Railroad, and strongly entrenched himself, and here properly began the siege of Atlanta. It became apparent almost immediately that he would attempt our left. He began to mass his forces in that quarter. On the 28th it became manifest that the enemy desired to place his left on Utoy creek. I desired to hold the Lick-Skillet road. and accordingly ordered Lieutenant General Lee, who on the 25th had relieved Major General Cheatham from the command of the corps formerly commanded by myself, to move his forces so as to prevent the enemy from gaining that road. He was ordered to hold the enemy in check on a line nearly parallel with the Lick-Skillet road, running through to Ezra Church. General Lee, finding that the enemy had already gained that position, engaged him with the intention to recover that line. This brought on the engagement of the 28th. General Stewart was ordered to support General Lee. The engagement continued until dark, the road remaining in our possession.

On the 27th July I received information that the enemy's cavalry was moving round our right with the design of interrupting our communication with Macon. The next day a large cavalry force also crossed the Chattahoochee river at Campbellton, moving round our left. Major General Wheeler was ordered to move upon the force on the right, while Brigadier General Jackson, with Hawson's and Ross's brigades, was sent to look after those moving on the left. I also dispatched Lewis's brigade of infantry down the Macon Railroad to a point about where they would probably strike the road. The force on the left succeeded in reaching the road, tearing up an inconsiderable part of the track. It was the design of the enemy to unite his forces at the railroad, but in this he was defeated. The movement was undertaken by the enemy on a grand scale, having carefully picked his men and horses. A Federal force, under General Stoneman, moved further south against Macon. He was defeated by our forces under Brigadier General Iverson. General Wheeler, leaving General Kelly to hold the force on the right, moved against that already at the railroad. He succeeded in forcing them to give battle near Newnan on the, 30th, and routed and captured or destroyed the whole force. Too much credit cannot be given General Wheeler for the energy and skill displayed. He captured two pieces of artillery, nine hundred and fifty (950) prisoners, and many horses, equipments, &c. Brigadier General Iverson captured two pieces of artillery and five hundred (500) prisoners. Believing the enemy's cavalry well broken, and feeling myself safe from any further serious operations of a like nature, I determined to dispatch a force of cavalry to the enemy's rear, with the hope of destroying his communications. I accordingly ordered Major General Wheeler, with four thousand five hundred (4500) cavalry, to effect this object. He succeeded in partially interrupting the enemy's communications by railroad. This still left sufficient cavalry to meet the necessities of the Army. This is sufficiently shown by the fact that several determined cavalry movements were subsequently attempted, and successfully met by our cavalry.

From this time till the 26th of August there is nothing of any particular moment to mention. The enemy gradually extended his right, and I was compelled to follow his movement; our entire front was covered with a most excellent abatis and other obstructions. Too much credit cannot be given the troops generally for the industry and endurance they displayed under the constant fire of the enemy.

On the 26th of August the enemy abandoned his works on the extreme right, and took up a line, the left resting in front of our works on the Dalton Railroad, and extending to the railroad crossing the river. Again he withdrew on the night of the 27th across Utoy creek, throwing one corps across the river to hold the railroad crossing and the intermediate points. His left then rested on the Chattahoochee river strongly fortified and extending across the West Point Railroad. The corps defending the crossing of the Chattahoochee, his works on this side of the river, and the obstacle formed by the Utoy and Camp creek rendered it impossible for me to attack him with any possibility of success, between the river and railroad.

On the 30th it became known that the enemy was moving on Jonesboro' with two corps. I determined, upon consulting with the corps commanders, to move two corps to Jonesboro' during the night, and to attack and drive the enemy at that place across Flint river. This I hoped would draw the attention of the enemy in that direction and that he would abandon his works on the left, so that I could attack him in flank. I remained in person with Stewart's Corps and the militia in Atlanta. Hardee's and Lee's Corps moved accordingly, Hardee in command. It was impressed upon General Hardee that the fate of Atlanta depended upon his success. Six hours before I had any information of the result of his attack, I ordered Lee to return in the direction of Atlanta, to be ready to commence the movement indicated in the event of success, and, if unsuccessful, to cover the evacuation of Atlanta which would thus be compelled. As it turned out unsuccessful, it allowed the enemy the opportunity either to strike us as we marched out of Atlanta, or to concentrate on Hardee. Lee's Corps constituted a guard against the former, and I did not fear the destruction of Hardee before Stewart and Lee could join him, as his position on a ridge between two rivers I thought strong in front, and want of time would prevent the enemy from attacking him in flank. The small loss in Hardee's Corps, and the much greater loss of the enemy, show my views to have been correct. The attack at Jonesboro' failed, though the number of men on our side considerably exceeded that of the enemy. The vigor of the attack may be in some sort imagined, when only fourteen hundred (1400) were killed and wounded out of the two corps engaged. The failure necessitated the evacuation of Atlanta. Thirty-four thousand (34,000) prisoners at Andersonville, Georgia, in my rear compelled me to place the Army between them and the enemy, thus preventing me at that time from moving on his communications and destroying his depots of supplies at Marietta. A raid of cavalry could easily have released those prisoners, and the Federal commander was prepared to furnish them arms; such a body of men, an Army of itself, could have overrun and devastated the country from West Georgia to Savannah. The subsequent removal of the prisoners, at my request, enabled me to make the movement on the enemy's communications at a later period.

On the night of the 1st of September we withdrew from Atlanta. A train of ordnance stores and some railroad stock had to be destroyed, in consequence of the gross neglect of the chief quarter master to obey the specific instructions given him touching their removal. He had ample time and means, and nothing whatever ought to have been lost.

On the 1st of September Hardee's Corps was attacked in position at Jonesboro.' The result was the loss of eight guns and some prisoners. Hardee then retired to Lovejoy's Station, where he was joined by Stewart's and Lee's Corps. The militia numbering about three thousand (3000), under Major General G. W. Smith, was ordered to Griffin. It is proper to remark here that this force rendered excellent and gallant service during the siege of Atlanta. The enemy followed and took position in our front. On the 6th of September, however, he abandoned his works and returned to Atlanta. Here properly ended the operations about Atlanta. Of the forces turned over to me nearly two months before, and since that day, daily engaged in battle and skirmishes, with a greatly superior enemy, there were remaining effective, as shown by the return of 20th September, infantry, twenty-seven thousand and ninety-four (27,094); cavalry, ten thousand five hundred and forty-three (10,543); artillery, two thousand seven hundred and sixty-six (2766). There had been sent to Mobile one brigade of infantry eight hundred (800) strong, and to Macon three battalions of artillery, eight hundred (800) strong. The militia had increased as stated, but counting it at the same as originally turned over, we have, against the aggregate turned over, forty-eight thousand seven hundred and fifty (48,750) present, forty thousand four hundred and three, sent off three thousand and one hundred (3100), making an aggregate of forty-three thousand five hundred and three (43,503,) thus giving a total loss of all arms of five thousand two hundred and forty-seven (5247) men.

A serious question was now presented to me. The enemy would not certainly long remain idle. He had it in his power to continue his march to the south, and force me to fall back on Alabama, for subsistence. I could not hope to hold my position. The country being a plain had not natural strength, nor was there any advantageous position upon which I could retire. Besides the *morale* of the Army, greatly improved during the operations around Atlanta, had again become impaired in consequence of the recurrence of retreat, and the Army itself decreasing in strength day by day. Something was absolutely demanded, and I rightly judged that an advance, at all promising success, would go far to restore its fighting spirit. Thus I determined on consultation with the corps commanders to turn the enemy's right flank and attempt to destroy his communications and force him to retire from Atlanta. The operations of the cavalry under Wheeler, in Georgia, and under Forrest, in Tennessee, proved to me conclusively and beyond a doubt that all the cavalry in the service could not permanently interrupt the railroad communications in the enemy's rear sufficiently to cause him to abandon his position. To accomplish anything, therefore, it became necessary for me to move with my whole force. Causing the iron to be removed from the several railroads out of Atlanta for distances of forty miles, and directing railroad stock to be restored to the West Point Railroad, the movement to the left toward that road began on the 18th of September. Arriving at that road the Army took position, with the left touching the Chattahoochee river, and covering that road where it remained several days to allow the accumulation of supplies at Blue Mountain and a sufficiency with which to continue the movement. On the 29th of September it left its bivouac near Palmetto, Georgia, with Jackson's cavalry in front, Brigadier General Iverson with his command being left in observation of the enemy in and around Atlanta, and moving first on the prolongation of its left flank to the westward it crossed the Chattahoochee river the same day on a pontoon bridge at Pumpkin Town and Phillips's Ferry, while our supplies which we brought by wagon from Newnan, Georgia, crossed at Moore's Ferry, where we had constructed a temporary trestle bridge. As soon as we crossed the river the Army moved at once to the immediate vicinity of Lost Mountain, reaching there on the 3d of October, our cavalry during the march watching the enemy on our front and right flank, and occasionally skirmishing with his cavalry along the banks of South Water creek. On the 4th of October Lieutenant General Stewart's Corps, in obedience to my orders, struck the enemy's railroad at Ackworth and Big Shanty, captured the garrisons at both places, consisting of some four hundred (400) prisoners, with some animals and stores.

Hearing that the enemy had a quantity of stores at Allatoona, I determined, if possible, to destroy the bridge over the Etowah river, and directed Lieutenant General Stewart to send a division also to Allatoona, instructing the officer in command to destroy the railroad there and take possession of the place, if, in his judgment, when he reached there, he deemed it practicable. Accordingly, Major General French was sent, who attacked the place early on the morning of the 6th of October, and quickly carried the enemy's outer line of works, drawing him into a redoubt, and with that exception carried the place. Just at this critical juncture he (General French) received information which he considered correct, but which subsequently proved false, that a large body of the enemy were moving to cut him off from the remainder of the Army, and he immediately withdrew his command from the place without having accomplished the desired object.

Lieutenant General Stewart's command succeeded in destroying completely some ten miles of the railroad. These operations caused the enemy to move his Army, except one corps, from Atlanta to Marietta, threatening an advance in the direction of our position at Lost Mountain; but not deeming our Army in condition for a general engagement I withdrew it, on the 6th of October, to the westward, continuing to march daily, and crossed the Coosa river near Coosaville, and moved up the west bank of Oostenaula, and striking the railroad again between Resaca and Mill Creek Gap, just above Dalton, on the 13th of October, destroying the railroad from Resaca to Tunnel Hill, capturing the enemy's posts at Tilton, Dalton, and Mill Creek Gap, with about one thousand (1000) prisoners and some stores. I again withdrew the Army from the railroad, moving from the southwest towards Gadsden, Alabama, the enemy following and skirmishing constantly with our cavalry, then under the command of Major General Wheeler, who had joined the Army on the march just before it crossed the Coosa river.

The Army reached Gadsden, Alabama, on the 20th of October, at which point General G. T. Beauregard, commanding the Military Division of the West, joined us. It had been my hope that my movements would have caused the enemy to divide his forces and that I might gain an opportunity to strike him in detail. This, however, he did not do. He held his entire force together in his pursuit, with the exception of the corps which he had left to garrison Atlanta. The *morale* of the Army had already improved, but upon consultation with my corps commanders, it was not thought to be yet in condition to hazard a general engagement while the enemy remained intact. I met at this place a thorough supply of shoes and other stores. I determined to cross the Tennessee river at or near Gunter's Landing, and strike the enemy's communications again near Bridgeport, force him to cross the river, also to obtain supplies and thus we should at least recover our lost territory. Orders had been sent by General Beauregard to General Forrest to move with his cavalry into Tennessee.

Unfortunately, however, these orders did not reach him in time. As I had not a sufficient cavalry force without his to protect my trains in Tennessee, I was compelled to delay the crossing and move further down the river to meet him. The Army arrived at Florence on the 31st of October. This unfortunate delay allowed the enemy time to repair the damage to his railroad and to accumulate at Atlanta sufficient supplies, to enable him to return the greater part of his Army to that place and move with it through to the Atlantic coast. The remainder he threw across the Tennessee under Thomas. When our Army arrived at Florence it had entirely recovered from the depression that frequent retreats had created. The enemy having for the first time divided his forces, I had to determine which of the two parts to direct my operations against. To follow the forces about, to move through Georgia under Sherman, would be to again abandon the regained territory to the forces under Thomas, with little hope of being able to reach the enemy in time to defeat his movement, and also to cause desertion and greatly impair the *morale* or fighting spirit of the Army by what would be considered a compulsory retreat.

I thought the alternative clear that I should move upon Thomas. If I succeeded in beating him, the effect of Sherman's movement would not be great, and I should gain in men sufficiently to compensate for the damages he might inflict. If beaten, I should leave the Army in better condition than it would be if I attempted a retrograde movement against Sherman.

Upon all these questions I had a full and free conference with General Beauregard at Tuscumbia. General Beauregard left it optional with me either to divide the Army, sending a part after Sherman, and to push on with the remainder, or to move forward at once against Thomas with the entire force. The Army I thought too small to divide. I so informed him, when he directed me by telegraph, to push forward at once. Forrest's cavalry joined me on the 21st of November, and the movement began, Major General Cheatham's Corps taking the road towards Waynesboro', and the other two corps moving on roads somewhat parallel with this, but more to the eastward with the cavalry under General Forrest, in the advance and upon their right flank. The enemy's forces at this time were concentrated at Pulaski, with some force also at Lawrenceburg. I hoped to be able to place the Army between these forces of the enemy and Nashville, but he evacuated Pulaski upon the 23d, hearing of our advance (our cavalry having furiously driven off their forces at Lawrenceburg), and moved rapidly by the turnpike and railroad to Columbia.

The want of a good map of the country, and the deep mud through which the Army marched, prevented our overtaking the enemy before he reached Columbia, but, on the evening of the 27th of November, our Army was placed in position in front of his works at that place. During the night, however, he evacuated the town, taking position on the opposite side of the river about a mile and a-half from the town, which was considered quite strong in front. Late in the evening of the 28th of November General Forrest with most of his command crossed Duck river a few miles above Columbia, and I followed, early in the morning of the 29th, with Stewart's and Cheatham's Corps, and Johnston's Division of Lee's Corps, leaving the other divisions of Lee's Corps in the enemy's front at Columbia. The troops moved in light marching order with only a battery to the corps, my object being to turn the enemy's flank by marching rapidly on roads parallel to the Columbia and Franklin pike at or near Spring Hill, and to cut off that portion of the enemy at or near Columbia. When I had gotten well on his flank the enemy discovered my intention, and began to retreat on the pike towards Spring Hill. The cavalry became engaged near that place about mid-day, but his trains were so strongly guarded that they were unable to break through them. About 4 p. m. our infantry forces—Major General Cheatham in the advance—commenced to come in contact with the enemy about two miles from Spring Hill, through which place the Columbia and Franklin pike runs. The enemy was at this time moving rapidly along the pike, with some of his troops formed on the flank of his column to protect it. Major General Cheatham was ordered to attack the enemy at once vigorously and get possession of this pike, and, although these orders were frequently and earnestly repeated, he made but a feeble and partial attack, failing to reach the point indicated. Had my instructions been carried out, there is no doubt that we should have possessed ourselves of this road. Stewart's Corps and Johnston's Division were arriving upon the field to support the attack. Though the golden opportunity had passed with daylight, I did not at dark abandon the hope of dealing the enemy a heavy blow. Accordingly, Lieutenant General Stewart was furnished a guide and ordered to move his corps beyond Cheatham's, and place it across the road beyond Spring Hill. Shortly after this General Cheatham came to my headquarters, and when I informed him of Stewart's movement, he said that Stewart ought to form on his right. I asked if that would throw Stewart across the pike. He replied that it would and a mile beyond. Accordingly, one of Cheatham's staff officers was sent to show Stewart where his (Cheatham's) right rested. In the dark and confusion he did not succeed in getting the position desired, but about 11 p. m. went into bivouac.

About 12 p. m., ascertaining that the enemy was moving in great confusion, artillery wagons and troops intermixed, I sent instructions to General Cheatham to advance a heavy line of skirmishers against him and still further impede and confuse his march. This was not accomplished. The enemy continued to move along the road in hurry and confusion within hearing nearly all the night. Thus was lost a great opportunity of striking the enemy for which we had labored so long, the greatest this campaign had offered, and one of the greatest during the war. Lieutenant General Lee, left in front of the enemy at Columbia, was instructed to press the enemy the moment he abandoned his position at that point. The enemy did not abandon his works at that place till dark, showing that his trains obstructed the road for fifteen miles during the day and a great part of the night. At daylight we followed as fast as possible towards Franklin; Lieutenant General Stewart in the advance, Major General Cheatham following, and General Lee, with the trains moving from Columbia on the same road. We pursued the enemy rapidly and compelled him to burn a number of his wagons. He made a feint as if to give battle on the hills about four miles south of Franklin, but, as soon as our forces began to deploy for the attack, and to flank him on his left, he retired slowly to Franklin.

I learned from dispatches captured at Spring Hill from Thomas to Schofield, that the latter was instructed to hold that place till the position at Franklin could be made secure, indicating the intention of Thomas to hold Franklin and his strong works at Murfreesboro'. Thus I knew that it was all important to attack Schofield before he could make himself strong, and if he should escape at Franklin, he would gain his works about Nashville. The nature of the position was such as to render it inexpedient to attempt any further flank movement, and I, therefore, determined to attack him in front and without delay. On the 30th of November Stewart's Corps was placed in position on the right, Cheatham's on the left, and the cavalry on either flank, the main body of the cavalry on the right under Forrest. Johnston's Division of Lee's Corps also became engaged on the left during the engagement. The line advanced at 4 p. m., with orders to drive the enemy into or across the Big Harpeth river, while General Forrest, if successful, was to cross the river and attack and destroy his trains and broken columns. The troops moved forward most gallantly to the attack. We carried the enemy's first line of hastily constructed works handsomely. We then advanced against his interior line, and succeeded in carrying it also in some places. Here the engagement was of the fiercest possible character. Our men possessed themselves of the exterior of the works, while the enemy held the interior. Many of our men were killed entirely inside the works. The brave men captured were taken inside his works in the edge of the town. The struggle lasted till near midnight, when the enemy abandoned his works and crossed the river, leaving his dead and wounded in our possession. Never did troops fight more gallantly. The works of the enemy were so hastily constructed that, while he had a slight abatis in front of a part of his line, there was none on his extreme right. During the day I was restrained from using my artillery on account of the women and children remaining in the town. At night it was massed ready to continue the action in the morning, but the enemy retired. We captured about a thousand prisoners and several stands of colors. Our loss in killed, wounded, and prisoners was four thousand five hundred (4500). Among the killed was Major General P. R. Cleburne, Brigadier Generals Gist, John Adams, Strahl, and Granberry. Major General Brown, Brigadier Generals Carter, Manigault, Quarles, Cockrell, and Scott were wounded, and Brigadier General Gordon captured. The number of dead left by the enemy on the field indicated that his loss was equal or near our own. The next morning at daylight, the wounded being cared for and the dead buried, we moved forward towards Nashville, Forrest with his cavalry pursuing the enemy vigorously.

On the 2d of December the Army took position in front of Nashville, about two miles from the city. Lieutenant General Lee's Corps constituted our centre, resting upon the Franklin pike, with Cheatham's Corps upon the right, and Stewart's on the left, and the cavalry on either flank extending to the river. I was causing strong detached works to be built to cover our flanks, intending to make them enclosed works, so as to defeat any attempt of the enemy should he undertake offensive movements against our flank and rear. The enemy still held Murfreesboro' with about six thousand (6000) men strongly fortified. He also held small forces at Chattanooga and Knoxville. It was apparent that he would soon have to take the offensive to relieve his garrisons at those points or cause them to be evacuated, in which case I hoped to capture the forces at Murfreesboro', and should then be able to open communication with Georgia and Virginia. Should he attack me in position, I felt that I could defeat him, and thus gain possession of Nashville with abundant supplies for the Army. This would give me possession of Tennessee. Necessary steps were taken to furnish the Army with supplies, which the people were ready and willing to furnish. Shoe shops were in operation in each brigade. We had captured sufficient railroad stock to use the road to Pulaski, and it was already in successful operation. Having possession of the State, we should have gained largely in recruits, and could at an early day have moved forward to the Ohio, which would have frustrated the plans of the enemy as developed in his campaign towards the Atlantic coast.

I had sent Major General Forrest, with the greatest part of his cavalry and Bates's Division of infantry to Murfreesboro', to ascertain if it was possible to take the place. After a careful examination and reconnoissance in force, in which I am sorry to say the infantry behaved badly, it was determined that nothing could be accomplished by assault. Bates's Division was then withdrawn, leaving Forrest with Jackson's and Buford's Divisions of cavalry in observation. Mercer and Palmer's brigades of infantry were sent to replace Bates's Division. Shortly afterwards Buford's Division was withdrawn and ordered to the right of the Army on the Cumberland river.

Nothing of importance occurred until the morning of the 15th of December, when the enemy, having received heavy reinforcements, attacked simultaneously both our flanks. On our right he was handsomely repulsed with heavy loss, but on our left, towards evening, he carried some partially completed redoubts of those before mentioned.

During the night of the 15th our whole line was shortened and strengthened; our left was also thrown back; dispositions were made to meet any renewed attack. The corps of Major General Cheatham was transferred from our right to our left, leaving Lieutenant General Lee on our right, who had been previously in the centre, and placing Lieutenant General Stewart's Corps in the centre, which had been previously the left. Early on the 16th of December the enemy made a general attack on our lines, accompanied by a heavy fire of artillery. All his assaults were repulsed with heavy loss till 3.30 p. m., when a portion of our line to the left of the centre occupied by Bates's Division suddenly gave way. Up to this time no battle ever progressed more favorably; the troops in excellent spirits, waving their colors and bidding defiance to the enemy. The position gained by the enemy being such as to enfilade our line caused in a few moments our entire line to give way, and our troops to retreat rapidly down the pike in the direction of Franklin, most of them, I regret to say, in great confusion, all efforts to re-form them being fruitless. Our loss in artillery was heavy—fifty-four guns. Thinking it impossible for the enemy to break our line, the horses were sent to the rear for safety, and the giving way of the line was so sudden that it was not possible to bring forward the horses to move the guns which had been placed in position. Our loss in killed and wounded was small.

At Brentwood, some four miles from our line of battle, the troops were somewhat collected, and Lieutenant General Lee took command of the rear guard, camping for the night in the vicinity. On leaving the field, I sent a staff officer to inform General Forrest of our defeat, and to direct him to rejoin the Army with as little delay as possible, to protect its rear, but owing to the swollen condition of the creeks, caused by the heavy rain then falling, he was unable to join us until we reached Columbia, with the exception of a portion of his command, which reached us while the enemy was moving from Franklin to Spring Hill. On the 17th we continued the retreat towards Columbia, camping for the night at Spring Hill. During this day's march the enemy's cavalry pressed with great boldness and activity, charging our infantry repeatedly with the sabre, and at times penetrating our lines. The country being open was favorable to their operations. I regret to say that also on this day Lieutenant General Lee, commanding the covering force, was severely wounded in the foot. We continued our retreat across Duck river to Columbia, the corps alternating as rear guards to the Army. Lieutenant General Lee and the corps commanded by him deserves great credit.

After the fight at Nashville I at first hoped to be able to remain in Tennessee, on the line of Duck river, but after arriving at Columbia I became convinced that the condition of the Army made it necessary to recross the Tennessee without delay, and on the 21st the Army resumed its march for Pulaski, leaving Major General Walthall with Ector's, Strahl's, Maney's, Granberry's and Palmer's infantry brigades at Columbia as a rear guard under General Forrest. From Pulaski I moved by the most direct road to the Bainbridge crossing on the Tennessee river, which was reached on the 25th, where the Army crossed without interruption, completing the crossing on the 27th, including our rear guard, which the enemy followed with all his cavalry and three corps of infantry to Pulaski, and with cavalry between Pulaski and the Tennessee river. After crossing the river the Army moved by easy marches to Tupelo, Mississippi. Our pontoon and supply trains were ordered at once to the vicinity of Columbus, Mississippi, by the most direct route, that the animals might be more easily foraged, and while on the march there were pursued by a small body of the enemy's cavalry, and owing to the neglect of Brigadier General Roddy's cavalry were overtaken, and the pontoon train and a small portion of the supply train destroyed.

Here finding so much dissatisfaction throughout the country, as in my judgment to greatly impair, if not destroy, my usefulness and counteract my exertions, and with no desire but to serve my country, I asked to be relieved, with the hope that another might be assigned to the command, who might do more than I could hope to accomplish. Accordingly, I was so relieved on the 23d of January by authority of the President.

And now, lest an opportunity should not be again presented, I trust I may be pardoned for noticing in self-defence one or two statements in General Johnston's report of the previous operations of this Army, which has just been given to the public, in which the action of Lieutenant General Polk and myself has been impugned. I thoroughly understand that it is not the part of an officer to state what may have occurred from time to time in council, but a charge publicly made ought certainly to be publicly met.

In General Johnston's report he says: "On the morning of the 19th (May), when half of the Federal Army was near Kingston, the two corps at Cassville were ordered to advance against the troops that had followed them from Adairsville, Hood leading on the right. When the corps had advanced some two miles, one of his staff officers reported to Lieutenant General Hood that the enemy was approaching on the Canton road, in rear of the right of our original position. He drew back his troops and formed them across that road. When it was discovered that the officer was mistaken, the opportunity had passed by the near approach of the Federal Army. Expecting to be attacked, I drew up my troops in what seemed to me an excellent position—a bold ridge immediately in rear of Cassville, with an open valley before it. The fire of the enemy's artillery commenced soon after the troops were formed, and continued until night. Soon after dark Lieutenant Generals Polk and Hood expressed to me decidedly the opinion formed upon the observation of the afternoon that the Federal Artillery would render their positions untenable the next day, and urged me to abandon the ground immediately and cross the Etowah. Lieutenant General Hardee, whose position I thought weakest, was confident he could hold it. The other two officers, however, were so earnest and so unwilling to depend upon the ability of their corps to defend the ground that I yielded, and the Army crossed the Etowah on the 28th, a step which I have regretted ever since." For myself and the good and great man now deceased, with whom I am associated in this stricture, I offer a statement of the facts in reply: After the Army had arrived at Cassville, I proposed to General Johnston, in the presence of Generals

Hardee and Polk, to move back upon the enemy and attack him at or near Adairsville, urging as a reason that our three corps could move back, each upon a separate road, while the enemy had but one main road upon which he could approach that place. No conclusion was obtained. While Generals Polk and Hardee and myself were riding from General Johnston's headquarters the matter was further discussed; General Polk enthusiastically advocated, and General Hardee also favoring the proposition. It was then suggested that we should return and still further urge the matter on General Johnston. We, however, concluded to delay till the morning. The next morning, while we were assembled at General Johnston's headquarters, it was reported that the enemy was driving in the cavalry on the Adairsville road, in front of Polk's position. Polk's Corps was in line of battle, and my corps was in bivouac on his right. We all rode to the right of Polk's line, in front of my bivouac. Hardee soon left and went to his position, which was on the left, there being some report of the enemy in that direction. General Johnston said to me, "You can, *if you desire*, move your corps to the Ironton road, and if Howard's Corps is there, you can attack it." My troops were put in motion; at the head of the column I moved over to this road, and found it in possession of our own dismounted cavalry, and no enemy there. While in motion, a body of the enemy, which I supposed to be cavalry, made its appearance on the Canton road in rear of the right of my original position. Major General Hindman was then in that direction with his division, to ascertain what force it was keeping the other two divisions in the vicinity of the Ironton road. *It was not a mistake*, as General Johnston states, that the force appeared, as is shown from the fact that Major General Hindman had men wounded from the small arms and artillery fired from this body. Major James Hamilton, of my staff, was sent to report to General Johnston the fact that the enemy had appeared on the Canton road. During Major Hamilton's absence Brigadier General Mackall, chief of staff, rode up in great haste and said that General Johnston directed that I should not separate myself so far from General Polk. I called his attention to where General Polk's right was resting, and informed him that I could easily form upon it, and orders were given to that effect, throwing back my right to look after this body, which turned out to be the enemy's cavalry. Feeling that I had done all which General Johnston had given me *liberty* to do, I then rode to his headquarters, where General Johnston decided to take up his line on the ridge in rear of the one occupied by General Polk, a line which was enfiladed by heights of which the enemy would at once possess himself, as was pointed out to General Johnston by Brigadier General Shoupe, commanding the artillery. In a very short time thereafter the enemy placed his artillery on these heights, and began to enfilade General Polk's line. Observing the effect upon the troops of this fire, I was convinced that the position was unsuited for defence. Accordingly General Polk and myself said to General Johnston that our positions would prove untenable for defence, but that we were in as good position to advance upon the enemy as could be desired. We told him that *if he did not intend to take the offensive* he had better change our position. He accordingly ordered the Army across the Etowah. It will thus be seen that I received *no order* to give battle, and I believe that had General Polk received such an order he would have mentioned it to me. Were General Polk now alive, he would be astounded at the accusation made against him. Again, General Johnston says, "that the usual skirmishing was kept up on the 28th (May); Lieutenant General Hood was instructed to put his corps in position during the night to attack the enemy's left flank at dawn the next morning, the rest of the Army to join in the attack successively from right to left. On the 29th (May), Lieutenant General Hood, finding the Federal left covered by a division which had entrenched itself in the night, thought it inexpedient to attack; so reported and asked for instructions. As the resulting delay made the attack inexpedient, even had it not been so before by preventing surprise upon which success in a great measure depended, he was recalled."

The enemy, on the 28th, had extended his left flank across Allatoona creek and along the Ackworth road. At my own suggestion General Johnston directed me to move my corps and strike the enemy's left. Upon arriving the next morning and while moving to accomplish this, I found that the enemy had retired his flank a mile and strongly fortified it. The opportunity having thus passed by the act of the enemy and not by my delay, I reported the fact to General Johnston, deeming it best that the attack should not be made, and the instructions to me were countermanded.

My operations are now fully stated. It may not be improper to close with a general *resumé* of the salient points presented. I was placed in command under the most trying circumstances which can surround an officer when assigned to a new and most important command. The Army was enfeebled in number and in spirit by long retreat and by severe and apparently fruitless losses. The Army of Tennessee between the 13th and 20th May, two months before, numbered seventy thousand (70,000) effective arms-bearing men, as the official reports show. It was at that time in most excellent condition and in full hope. It had dwindled day by day in partial engagements and skirmishes without an action that could properly be called a battle, to forty-seven thousand two hundred and fifty (47,250), exclusive of one thousand five hundred militia which joined in the interim. What with this constant digging and retreating from Dalton to Atlanta, the spirit of the Army was greatly impaired and hope had almost left it. With this Army I immediately engaged the enemy, and the tone constantly improved and hope returned. I defended Atlanta, a place without natural advantages, or rather with all the advantages in favor of the enemy, for forty-three days. No point, of all passed over from Dalton down, was less susceptible of defence by nature. Every preparation was made for retreat. The Army lay in bivouac a short distance from the town, without attempting to construct works of defence in front of the camps, ready to resume the line of march as soon as the enemy pressed forward. I venture the statement that there was neither soldier nor officer in that Army who believed that in the open plain between Atlanta and the river a battle would be offered, which had so often been refused in strong positions on the mountains.

My first care was to make an entrenched line, and the enemy despairing of success in front threw his Army to the left and rear, a thing that he never could have done had it not been for the immense advantage the Chattahoochee river gave him. I arrived at Lovejoy's Station, having fought four battles, and the official reports of the Army on the 20th of September show an effective total of forty thousand four hundred and three (40,403) present, giving a total loss in all this time of five thousand two hundred and forty-seven (5247) men.

My reasons for undertaking the movement into Tennessee have, I think, been sufficiently stated already. Had I not made the movement, I am fully persuaded that Sherman would have been upon General Lee's communications in October, instead of at this time.

From Palmetto to Spring Hill the campaign was all that I could have desired. The fruits ought to have been gathered at that point. At Nashville, had it not have been for an unfortunate event, which could not justly have been anticipated, I think we would have gained a complete victory.

At any time it was in the power of the Army to retire from Tennessee in the event of failure, as is established by the leisurely retreat which was made under the most difficult and embarrassing circumstances.

It is my firm conviction that, notwithstanding that disaster, I left the Army in better spirits and with more confidence in itself, than it had at the opening of the campaign. The official records will show that my losses, including prisoners, during the entire campaign do not exceed ten thousand (10,000) men. Were I again placed in such circumstances, I should make the same marches and fight the same battles, trusting that the same unforeseen and unavoidable accident would not again occur to change into disaster a victory which had been already won.

In support of the statement touching the strength and losses of the Army, I respectfully tender the official records of the assistant adjutant general (Major Kinloch Falconer) alike on duty with General Johnston and myself.

Those who have seen much service in the field during this war will at once understand why it was that desertion, which had been so frequent on the retreat from Dalton to Atlanta, almost entirely ceased as soon as the Army assumed the offensive and took a step forward. I did not know of a desertion on the march from Palmetto to Dalton, or from Dalton to Florence. I am informed that the provost marshal general of the Army of Tennessee reports less than three hundred (300) desertions during the whole Tennessee campaign. The Tennessee troops entered the State with high hopes as they approached their homes. When the fortunes of war were against us, the same faithful soldiers remained true to their flag, and with rare exceptions followed it in retreat as they had borne it in advance.

But few of the subordinate reports have reached me. I am consequently unable, without risk of injustice, to describe the instances of individual skill and gallantry.

I invite special attention to the report of Major General G. W. Smith of the operations of the Georgia militia in the vicinity of Atlanta, the reports of Lieutenant General Stewart and his subordinate officers, herewith submitted. Maps of the campaign accompany this report.

Respectfully your obedient servant,

(Signed) J. B. HOOD, *General.*

OFFICIAL REPORT LIEUTENANT GENERAL S. D. LEE.

COLUMBUS, MISS., *January 30th, 1864.*

COLONEL:—Owing to my temporary absence from the Army, and to the movements of the troops, it will be impracticable to procure detailed reports from my subordinate officers, and I cannot, therefore, make a full report of the operations of my command during the recent campaign, but deem it proper to offer this until one more complete may be substituted.

I assumed command of "Wood's old Corps," consisting of Stevenson's, Clayton's, and Hindman's Divisions, the latter commanded by Brigadier General John C. Brown, on July 27th, 1864. The Army was then in position and entrenched around Atlanta, daily shifting its position to meet the flank movements of the enemy. On the 27th, Hindman's and Clayton's Divisions were withdrawn from the trenches and massed on the Lick-Skillet road. On the 28th, about 11 a. m., I received orders to move out on the Lick-Skillet road and check the enemy, who was then moving to our left, as it was desirable to hold that road, to be used for a contemplated movement. I soon found that the enemy had gained the road, and was gradually driving back our cavalry. Brown's Division was formed at once on the left of, and obliquely to the road, and Clayton's Division on the right, connecting by a line of skirmishers, with the main works around the city. As soon as Brown was formed he moved forward handsomely, drawing the enemy across the road, and to a distance one-half mile beyond, where he encountered temporary breastworks, from which he was driven back with considerable loss. Clayton's Division moved forward as soon as formed, about ten minutes after Brown's advance, and met with similar results. I found it difficult to rally Brown's Division, and move it against the enemy a second time. The consequence was, that one or two brigades of this division, as also of Clayton's Division, sustained heavy losses, because of the failure in the attack of portions of their lines. Walthall's Division, of Stewart's Corps, had moved out on the Lick-Skillet road, while Brown's and Clayton's Divisions were engaging the enemy. At my suggestion this division was thrown against the enemy when Brown had attacked. The enemy was still within easy range of the Lick-Skillet road, and I believed that he would yield before a vigorous attack. The effect, however, was a failure, and the troops were formed on the road, and during the night were withdrawn, by order of the Commanding General, to a more suitable position, connecting with the works immediately around Atlanta. The enemy had two corps engaged in this affair; still I am convinced that if all the troops had displayed equal spirit, we would have been successful, as the enemy's works were slight, and besides they had scarcely gotten into position when we made the attack. From the 28th July to the 5th of August the enemy cautiously pushed forward his lines towards ours, erecting new lines of works as he advanced. Several severe attacks were made upon the works of my skirmish line, but no assault was made on the main entrenched line. The enemy, in almost every instance, was severely repulsed.

On the 6th Major General Bates's Division of Hardee's Corps, which had reported to me temporarily in place of Stevenson's Division, which had been detached from my corps and placed in the works immediately in front of Atlanta, took position on my left, almost perpendicularly to our main line, and along the Sandtown road. This division constructed in one night a very strong skirmish line, and with such little display that the enemy, on the 6th, finding, as he supposed, only a slight impediment to the extension of his lines, at once moved a corps to the attack, which was signally and handsomely repulsed. Much credit is due General Bates and his division for their conduct. The enemy was exceedingly cautious in his movements after this affair. His extension to our left was gradual, and he seemed determined to push his lines more closely to ours, in my front, with a view of making an assault. The skirmishing along Patton Anderson's—formerly Hindman's—and Clayton's Divisions, amounted to almost an engagement for a week. Hardee's Corps had been placed on my left to check the enemy, who continued extending to the left. The enemy, about the 26th, retired his right flank from my front, making quite a detour to the left of my corps, which extended to the West Point and Atlanta Railroad, three-quarters of a mile beyond East Point. Stevenson's Division reported to me by 11 a. m. on the 30th of August. Hardee's Corps was on my left, and was gradually relieved by my corps, in order that it might extend further to our left. About 4 p. m. on the 30th I was notified that General Hardee would probably move to Jonesboro', and that it was desired that my corps should follow and support him. At Army headquarters, in Atlanta, about 9 p. m., it was decided that the column of the enemy, which was marching on Jonesboro' from the direction of the Atlanta and West Point Railroad, should be attacked early on the morning of the 31st, and crushed, if practicable, and that Lieutenant General Hardee, with his corps and my own, should be charged with the expedition. According to my recollection, the column of the enemy marching on Jonesboro' was the only body well defined and in motion, and that it consisted of about three Army Corps. I accordingly reported to General Hardee at General

Hood's headquarters; was advised that General Hardee's Corps, the left of which rested at Rough and Ready, four miles below East Point, on the Macon Railroad, commenced moving about 4 p. m. Orders were extended for my corps to move immediately after General Hardee's. The rear of Hardee's Corps was in motion about 11½ p. m. My corps was well closed up to it, and immediately following. Our progress was very slow, and the head of my column did not reach Rough and Ready till daylight. I ascertained that the delay was caused by a part of Hardee's Corps encountering the enemy about 12 p. m., August 30th, on the road on which they were moving, which made it necessary for the line of march to be changed to a neighborhood road. In consequence of this delay my corps did not arrive at Jonesboro' till near 10 a. m. on the 31st, but it reached there immediately in rear of General Hardee's last division. The last three brigades of my corps, in consequence of the distance they had marched, and having been on picket, arrived about 1.30 p. m.

The enemy had during the previous evening and night, effected a crossing of the Flint river, and made a lodgement on the east bank. The preliminaries for the attack were arranged. My corps was formed almost parallel to the railroad, immediately to the right of Jonesboro', connecting with Hardee's right, his line extending towards Flint river, and making almost a right angle with the railroad. It was found that Hardee's Corps did not cover as much ground as was expected, and I was instructed to extend my troops so as to fill up the interval; and my command was moved almost two divisions front to the left. The instructions given me were to attack as soon as Cleburne, who commanded Hardee's Corps, should become hotly engaged, he being ordered to swing to his right, and my corps to advance directly against the enemy, and if possible, swing to the left. The firing to my left on Cleburne's line, did not indicate a serious engagement, until the right division of Hardee's Corps became engaged. Being satisfied that the battle had commenced in earnest, I at once gave orders to my corps to move against the enemy. The attack was not made by the troops with that spirit and inflexible determination that would ensure success. Several brigades behaved with great gallantry, and in each brigade many instances of gallant conduct were exhibited by regiments and individuals, but generally the troops faltered in the charge, when they were much exposed and within easy range of the enemy's musketry, and when they could do but little damage to the enemy behind his works, instead of moving directly and promptly forward against the temporary and informidable works in their front. The attack was a feeble one, and a failure, with a loss to my corps of about thirteen hundred (1300) men in killed and wounded. The enemy being behind works, and apparently no impression having been made upon him by the attack on my left, where his line was supposed to be weakest, and Brigadier General Ross commanding a cavalry brigade on my immediate right, having reported the enemy moving to my right, I was induced not to renew the attack. During the night of the 31st, about 1 p. m., I received an order from Lieutenant General Hardee to march at once to Atlanta. My corps was immediately put in motion, and was halted by Major General M. S. Smith, chief engineer of the Army, about six miles from Atlanta, and there put in position to cover the evacuation of the city. On the morning of September 1st, I was ordered to move my command towards Lovejoy's Station, which place I reached on the 3d. The Army remained at Lovejoy till September 18th, when it commenced moving to Palmetto Station, on the Atlanta and West Point Railroad, where it arrived on the 19th.

Not having received the reports of my division commanders, it is impossible to notice those officers and commands deserving especial mention. It is my purpose to refer to their gallant deeds in a subsequent and more detailed report. I received at all times the cordial support of my division commanders—Major Generals Stevenson, Clayton and Brown, and afterwards Patton Anderson, commanding Hindman's old division, they always displayed great gallantry and zeal in time of battle.* I take pleasure in making a special mention of the gallantry of Brigadier General (now Major General) John C. Brown, during the engagement of the 28th on the Lick-Skillet road, and of Major General Stevenson and Clayton during the battle of Jonesboro' on August 31st. The officers of my personal staff, as also of the corps staff, behaved at all times with gallantry, and were energetic in the discharge of their duties.

I have the honor to be, yours respectfully,

S. D. LEE, *Lieutenant General.*

COLUMBUS, MISS., *January 30th, 1864.*

COLONEL:—I have the honor to offer the following as my official report of the operations of my corps, during the offensive movements commencing at Palmetto Station, Georgia, September 29th, 1864. It is impracticable, now in consequence of the movement of troops and my temporary absence from the Army, to obtain detailed reports from my division commanders.

* I regret to state that Major General Patton Anderson and Brigadier General Cummings were severely wounded in the action of the 31st, while nobly leading their troops against the enemy's works, and their services were lost to us during the remainder of the campaign.

As a corps commander, I regarded the *morale* of the Army greatly impaired after the fall of Atlanta, and in fact *before* its fall, the troops were not by any means in good spirits. It was my observation and belief that the majority of the officers and men were so impressed with the idea of their inability to carry even temporary breastworks, that when orders were given for attack, and there was a probability of encountering works, they regarded it as recklessness in the extreme. Being impressed with these convictions, they did not generally move to the attack with that spirit which nearly always ensures success. Whenever the enemy changed his position temporary works could be improvised in less than two hours, and he could never be caught without them. In making these observations, it is due to many gallant officers and commands to state that there were noticeable exceptions; but the feeling was so general that anything like a general attack was paralyzed by it. The Army having constantly yielded to the flank movements of the enemy, which he could make with but little difficulty by reason of his vastly superior numbers, and having failed in the offensive movements prior to the fall of Atlanta, its efficiency for further retarding the progress of the enemy was much impaired, and besides the advantages in the topography of the country south of Atlanta were much more favorable to the enemy for the movements of his superior numbers than the rough and mountainous country already yielded to him. In view of these facts, it was my opinion that the Army should take up the offensive, with the hope that favorable opportunities would be offered for striking the enemy successfully, thus ensuring the efficiency of the Army for future operations. These opinions were freely expressed to the Commanding General.

My corps crossed the Chattahoochee river on September 29th, and on October 3d, 1864, took position near Lost Mountain to cover the movements of Stewart's Corps on the railroad at Big Shanty and Allatoona. On October 6th I left my position near Lost Mountain, marching via Dallas and Cedar Town, crossing the Coosa river at Coosaville, October 10th, and moved on Resaca, partially investing the place by 4 p. m., on October 12th. The surrender of the place was demanded in a written communication, which was in my possession, signed by General Hood. The commanding officer refused to surrender, as he could have easily escaped from the forts with his forces, and crossed the Oostenaula river; I did not deem it prudent to assault the works, which were strong and well manned, believing that our loss would have been severe. The main object of appearing before Resaca being accomplished, and finding that Sherman's main Army was moving from the direction of Rome and Adairsville towards Resaca, I withdrew from before the place to Snake Creek Gap about mid-day on the 13th. The enemy made his appearance at the Gap on the 14th in large force, and on the 15th it was evident that his force amounted to several corps. Several severe skirmishes took place on the 15th, in which Deas's and Bradley's brigades of Johnson's Division were principally engaged.

This Gap was held by my command until the balance of the Army had passed through Mattox's Gap, when I followed with the corps through the latter. The Army moved to Gadsden, where my corps arrived on October 21st. At this point clothing was issued to the troops, and the Army commenced its march towards the Tennessee. My corps reached the vicinity of Leighton, in the Tennessee Valley, October 29th. Stewart's and Cheatham's Corps were then in front of Decatur. On the night of the 29th I received orders to cross the Tennessee river at Florence, Alabama. By means of the pontoon boats, two brigades of Johnson's Division were thrown across the river two and one-half miles above South Florence, and Gibson's brigade of Clayton's Division was crossed at South Florence. The enemy occupied Florence with about one thousand (1000) cavalry, and had a strong picket at the railroad bridge. The crossing at this point was handsomely executed, and with much spirit by Gibson, under the direction of General Clayton, under cover of several batteries of artillery. The distance across the river was about one thousand yards. The troops landed, and after forming, charged the enemy, and drove him from Florence. The crossing was spirited, and reflected much credit on all engaged in it. Major General Ed. Johnson experienced considerable trouble in crossing his two brigades because of the extreme difficulty of managing the boats in the shoals. He moved from the north bank of the river late in the evening with one brigade, Sharpe's Mississippi, and encountered the enemy on the Florence and Huntsville road about dark. A spirited affair took place, in which the enemy were defeated, with a loss of about forty (40) killed, wounded, and prisoners. The enemy retreated during the night to Shoal creek, about nine miles distant. The remainder of Johnson's and Clayton's Divisions were crossed on the night of the 30th, and on the morning of the 31st. Stevenson's Division was crossed on November 2d. My corps remained at Florence till November 20th, when the Army commenced moving for Tennessee, my command leading the advance, and marching in the direction of Columbia via Henryville and Mount Pleasant. I arrived in front of Columbia on the 26th, relieving Forrest's cavalry then in position there, which had followed the enemy from Pulaski. The force of the enemy occupying Columbia was two corps. They confined themselves to the main works around the city and their outposts, and skirmishers were readily driven in. On the night of the 27th the enemy evacuated Columbia, and crossed Duck river; Stevenson's Division of my corps entered the town before daylight. After crossing, the enemy took a strong position on the opposite side of the river, and entrenched, his skirmishers occupying rifle pits, two hundred and fifty yards from the river. There was considerable skirmishing across the river during the day, and some artillery firing, resulting in nothing of importance. On the morning of the 29th Johnson's Division of my corps was detached and ordered to report to the General Commanding. I was directed to engage and occupy the enemy near Columbia, while the other two corps and Johnson's Division would be crossed above, and moved to the rear of the enemy in the direction of Spring Hill. The entire force of the enemy was in front of Columbia till about mid-day on the 29th, when one corps commenced moving off, the other remaining in position as long as they could be seen by us, or even till dark. I had several batteries of artillery put in position to drive the skirmishers of the enemy from the vicinity of the river bank, and made a display of pontoons, running several of them down to the river under heavy artillery and musketry fire. Having succeeded in putting a boat in the river, Pettus's brigade of Stevenson's Division was thrown across, under the immediate direction of Major General Stevenson, and made a most gallant charge on the rifle pits of the enemy, driving a much superior force, and capturing the pits. The bridge was at once laid down, and the crossing commenced. The enemy left my front about 2½ a. m., on the morning of the 30th, and pursuit was made as rapidly as was prudent in the night time. The advance of Clayton's Division arrived at Spring Hill about 9 a. m., when it was discovered that the enemy had made his escape, passing around that portion of the Army in that vicinity. My corps, including Johnson's Division, followed immediately after Cheatham's Corps towards Franklin. I arrived near Franklin about 4 p. m.

The Commanding General was just about attacking the enemy with Stewart's and Cheatham's Corps, and he directed me to place Johnson's Division, and afterwards Clayton's, in position to support the attack. Johnson moved in rear of Cheatham's Corps, and finding that the battle was stubborn, General Hood instructed me to move forward in person; to communicate with General Cheatham, and if necessary, to put Johnson's Division in the fight. I met General Cheatham about dark, and was informed by him that assistance was needed at once. Johnson was immediately moved forward to the attack, but owing to the darkness and want of information as to the locality his attack was not felt by the enemy till about one hour after dark. This division moved against the enemy's breastworks under a heavy fire of artillery and musketry, gallantly driving the enemy from portions of his line. The brigades of Sharpe and Bradley (Mississippians), and of Deas (Alabamians), particularly distinguished themselves. Their dead were mostly in the trenches and on the works of the enemy, where they nobly fell in a desperate hand-to-hand conflict. Sharpe captured three stands of colors. Bradley was exposed to a severe enfilade fire, yet these noble brigades never faltered in the terrible night struggle. I have never seen greater evidences of gallantry than was displayed by this division, under command of that admirable and gallant soldier, Major General Edward Johnson. The enemy fought gallantly and obstinately at Franklin, and the position he held was, for infantry defence, one of the best I have ever seen. The enemy evacuated Franklin hastily during the night of the 30th. My corps commenced the pursuit about 1 p. m., on December 1st, and arrived near Nashville, about 2 p. m., on the 2d.

The enemy had then occupied the works around the city. My command was the centre of the Army in front of Nashville; Cheatham's Corps being on my right, and Stewart's on my left. Nothing of importance occurred until the 15th. The enemy was engaged in entrenching and strengthening its position. On the 15th the enemy moved out on our left, and a severe engagement was soon commenced. In my immediate front the enemy still kept up his skirmish line, though it was evident that his main force had been moved. My line was much extended, the greater part of my command being in single rank. About 12 m. I was instructed to assist Lieutenant General Stewart, and I commenced withdrawing troops from my line to send to his support. I sent him Johnson's entire division, each brigade starting as it was disengaged from the works. A short time before sunset the enemy succeeded in turning General Stewart's position, and a part of my line was necessarily changed to conform to his new line. During the night, Cheatham's Corps was withdrawn from my right and moved to the extreme left of the Army The Army then took position about one mile in rear of its original line, my corps being on the extreme right. I was instructed by the Commanding General to cover and hold the Franklin pike. Clayton's Division occupied my right, Stevenson's my centre, and Johnson's my left. It was evident, soon after daylight, that a large force of the enemy was being concentrated in my front on the Franklin pike. About 9 o'clock a. m., on the 16th, the enemy having placed a large number of guns in position opened a terrible artillery fire on my line, principally on the Franklin pike. This lasted about two hours, when the enemy moved to the assault. They came up in several lines of battle. My men reserved their fire till they were within easy range, and then delivered it with terrible effect. The assault was easily repulsed. It was renewed, however, several times, with spirit, but only to meet each time with a like result. They approached to within thirty yards of our line and their loss was very severe. Their last assault was made about 3½ p. m., when they were driven back in great disorder.

The assaults were made principally in front of Holtzclaus' (Alabama), Gibson's (Louisiana), and Stovall's (Georgia) brigades, of Clayton's Division, and Pettus's Alabama brigade of Stevenson's Division, and too much credit cannot be awarded Major General Clayton and these gallant troops for their conspicuous and soldierly conduct. The enemy made a considerable display of force on my extreme right during the day, evidently with the intention of attempting to turn our right flank. He made, however, but one feeble effort to use this force, when it was readily repulsed by Stovall's Georgia, and Bradley's Mississippi brigades, which latter had been moved to the right. Smith's Division, of Cheatham's Corps, reported to me about 2 p. m., to meet any attempt of the enemy to turn our right flank. It was put in position, but was not needed, and by order of the Commanding General it started to Brentwood, about 3½ p. m.

The artillery fire of the enemy during the entire day was very heavy, and right nobly did the artillery of my corps, under Lieutenant General Hoxton, perform their duty. Courtney's battalion, under Captain Douglass, was in Johnson's front; Johnson's battalion was in Stevenson's front, and Eldrige's battalion, under Captain Fenner, was in Clayton's front. The officers and men of the artillery behaved admirably, and too much praise cannot be bestowed upon this efficient arm of the service in the Army of Tennessee. The troops of my entire line were in fine spirits and confident of success (so much so that the men could scarcely be prevented from leaving their trenches to follow the enemy on and near the Franklin pike). But suddenly all eyes were turned to the centre of our line of battle, near the Granny White pike, where it was evident the enemy had made an entrance, although but little firing had been heard in that direction. Our men were flying to the rear in the wildest confusion, and the enemy following with enthusiastic cheers. The enemy at once closed towards the gap in our line and commenced charging on the left division—Johnson's—of my corps; but were handsomely driven back. The enemy soon gained our rear, and were moving on my left flank, when our line gradually gave way. My troops left their lines in some disorder, but were soon rallied and presented a good front to the enemy. It was a fortunate circumstance that the enemy was too much crippled to pursue us on the Franklin pike. The only pursuit made at that time was by a small force coming from the Granny White pike.

Having been informed by an aid of the General Commanding that the enemy was near Brentwood, and that it was necessary to get beyond that point at once, everything was hastened to the rear. When Brentwood was passed, the enemy was only one-half mile from the Franklin pike, when Chalmer's Cavalry was fighting them. Being charged with covering the retreat of the Army I remained in rear with Clayton's and part of Stevenson's Division, and halted the rear guard about seven miles north of Franklin, about 10 p. m., on the 16th. Early on the morning of the 17th our cavalry was driven in—in confusion—by the enemy; who at once commenced a most vigorous pursuit, his cavalry charging at every opportunity and in the most daring manner. It was apparent that they were determined to make the retreat a rout, if possible. Their boldness was soon checked by many of them being killed and captured by Pettus's Alabama, and Stovall's Georgia, brigades, with Bledsoe's battery, under Major General Clayton. Several guidons were captured in one of their charges. I was soon compelled to withdraw rapidly towards Franklin, as the enemy was throwing a force in my rear from both the right and left of the pike, on roads coming into the pike near Franklin, and five miles in my rear. This force was checked by Brigadier General Gibson, with his brigade, and a regiment of Buford's cavalry, under Colonel Shacklet.

The resistance which the enemy had met with early in the morning, and which materially checked his movements, enabled us to reach Franklin with but little difficulty. There the enemy appeared in considerable force and exhibited great boldness; but he was repulsed, and the crossing of the Harpeth river effected. I found that there was in the town of Franklin a large number of our own and of the enemy's wounded, and not wishing to subject them and the town to the fire of the enemy's artillery, the place was yielded with but little resistance. Some four or five hours were gained by checking the enemy about one and one-half miles south of Franklin, and by the destruction of the trestle bridge over the Harpeth, which was effected by Captain Coleman, the engineer officer on my staff, and a party of pioneers, under a heavy fire of the enemy's sharpshooters. About 4 p. m. the enemy, having crossed a considerable force, commenced a bold and vigorous attack, charging with his cavalry on our flanks and pushing forward his lines in our front. A more persistent effort was never made to rout the rear guard of a retiring column. This desperate attack was kept up till long after dark, but gallantly did the rear guard, consisting of Pettus's Alabama and Cummings's Georgia brigades, the latter commanded by General Watkins, of Stevenson's Division, and under that gallant and meritorious officer, Major General C. L. Stevenson, repulse every attack. Brigadier General Chalmers with his division of cavalry covered our flanks. The cavalry of the enemy succeeded in getting in Stevenson's rear, and attacked Major General Clayton's Division about dark; but they were handsomely repulsed, Gibson and Stovall's brigades being principally engaged. Some four or five guidons were captured from the enemy during the evening.

About 1 p. m. I was wounded while with the rear guard, but did not relinquish command of my corps till dark. Most of the details in conducting the returns from that time were arranged and executed by Major General C. L. Stevenson, to whom the Army is much indebted for his skill and gallant conduct during the day. I cannot close this report without alluding particularly to the conduct of the artillery of my corps on the 16th. Sixteen guns were lost on the lines. The greater portion of them were without horses, they having been disabled during the day. Many of the carriages were disabled also. The noble gunners, reluctant to leave their guns, fought the enemy, in many instances, till they were almost within reach of the guns. Major General Ed. Johnson was captured on the 16th. Being on foot he was unable to make his escape from the enemy in consequence of an old wound. He held his line as long as it was practicable to do so. The Army of Tennessee has sustained no greater loss than that of this gallant and accomplished soldier. To my division commanders—Stevenson, Johnson and Clayton—I am indebted for the most valuable services. They were always zealous in discharge of their duties. Although it is my desire to do so, I cannot now allude to the many conspicuous acts of gallantry exhibited by general, field, and company officers, and by the different commands. It is my intention to do so in future, when detailed reports are received. To the officers of my personal staff, and also of the corps staff, I am indebted for valuable services. They were always at their posts and ready to respond to the call of duty.

I have the honor to be yours, respectfully,

S. D. LEE, *Lieutenant General.*

NOTES.—Brigadier General Manigault, commanding a brigade of Alabamians and South Carolinians, was severely wounded in the engagement while gallantly leading his troops to the fight, and his two successors in command—Colonel Shaw was killed, and Colonel Davis wounded. During the affair around Columbia, the gallant and accomplished soldier, Colonel R. F. Beckham, commanding the artillery regiment of my corps, was mortally wounded, while industriously and fearlessly directing the artillery firing against the enemy. He was of the truest and best officers in the service.

S. D. LEE, *Lieutenant General.*

COLUMBUS, MISS., January 30th, 1864.

Report of the operations of Lee's Corps from the commencement of offensive operations at Palmetto, Georgia.

OFFICIAL REPORT OF LIEUTENANT GENERAL STEWART, OF OPERATIONS UNDER GENERAL HOOD.

HEADQUARTERS, STEWART'S CORPS, A. OF T.
NEAR TUPELO, MISS., *January 12th, 1865.*

COL. A. P. MASON, *A. A. General.*

SIR:—In compliance with the wishes of the Commanding General, I submit the following brief report of the operations of this corps from July 18th, 1864, the day on which General Hood took command of the Army of Tennessee, to September 29th, 1864, the day on which we re-crossed the Chattahoochee. On the 18th of July we lay in bivouac on the south side of Peach Tree creek, between the Marietta and Pace's Ferry road. On that or the following day we commenced entrenching, the enemy having crossed the Chattahoochee, and advanced, on Sunday, the 17th, to the vicinity of Peach Tree creek. This corps was on the left, Hardee's in the centre, Cheatham's, formerly Hood's, on the right of the Army. On the morning of the 20th it was decided at Army headquarters that at 1 o'clock, p. m. that day an attack should be made on the enemy by this corps and Hardee's. The plan was for the divisions (commencing on Hardee's right) to move forward successively *en echelon*, at intervals of some two hundred yards, to attack the enemy, drive him back to the creek, and then pass down the creek to the left. Should the enemy be found entrenched, his works were to be carried, every thing on our side of the creek was to be taken, and our crossing to the other side of the creek was to depend on our success. Such were the instructions of the Commanding General to General Hardee and myself. I was to hold a division in reserve. It seems that a division had been withdrawn from the lines on the right of Hardee's Corps. His corps and mine were to close to the right far enough to cover the space vacated by this division —the space to be divided between the two corps. This would have shifted my line a half division front to right, perhaps at most half a mile.

The division and brigade commanders were notified of the work to be done, and directed to reconnoitre the ground in front of their lines as far as practicable.

At 1 o'clock I found the left of Hardee's Corps just beginning to shift to the right. Feeling that this change was not important, and that not a moment was to be lost in making the attack contemplated, a staff officer was dispatched to the Commanding General to inform him of the fact, and requesting an order to stop the movement to the right and commence a forward movement. The result was, however, that to keep up connection with the other corps my line was moved a mile and a-half or two miles to the right, and my right division (Loring's) did not move forward—following the one on its right in the prescribed order, until nearly 4 o'clock. My instructions to division commanders, and through them to brigade and regimental commanders, were to move forward and attack the enemy; if found entrenched, to fix bayonets and carry his works; to drive him back to the creek, and then press down the creek; that we were to carry everything in our front on our side of the creek. Loring's Division was on the right, Walthall's in the centre, and French's, the reserve division, on the left.

The instructions given were obeyed promptly and with alacrity. Loring's Division moved forward and carried the works in their front, but were compelled to fall back by an enfilade fire from the right, because the left division of the other corps had not moved up to the attack. Walthall's Division also engaged the enemy with great spirit, while French was moved so as to cover the left, and be in supporting distance. Learning the cause of the check to Loring's and Walthall's Divisions, an officer was dispatched to request General Hardee to allow his left division to co-operate with Loring's in carrying the line in its front. Before an answer was received, a staff officer brought me an order from the Commanding General to retire to the entrenched line from which we had advanced, and the conflict terminated. The loss in Loring's and Walthall's Divisions, especially the former, was heavy. These commanders, their officers and men, behaved entirely to my satisfaction, and I cannot but think had the plan of the battle, as I understood it, been carried out fully, we would have achieved a great success. I beg leave to refer to subordinate reports, for the names of those who distinguished themselves by special gallantry, and of the brave officers and men whose blood was so freely, and it would seem so uselessly, shed on this occasion, as well as for those details it is impossible to embrace in this sketch.

Thursday night, July 21st, the Army fell back to the line around Atlanta, this corps occupying the space from Peach Tree creek to a redoubt on a hill to the left of the Turner's Ferry road, being still the left of the Army.

On the 22d we were again ordered to be in readiness to attack the enemy, following the corps on our right, but for reasons unknown to me the battle did not become general on that day. On the 28th the enemy, by extending to his right, had nearly gained the Lick-Skillet road, Loring's and Walthall's Divisions had been relieved at the trenches, and it was expected that French's would be that night. As I understood the instructions, General Lee, commanding corps, was to move out on the Lick-Skillet road, attack the enemy's right flank, and drive him from that road and the one leading from it by Mount Ezra Church. My own orders were to move with the divisions named to the point where our own line of works crossed the Lick-Skillet road. French's Division, when relieved, and one from some other corps, were to rejoin us, and at an early hour next morning we were to move out upon that road, turn to the right, pass in rear of the enemy, and attack. On reaching the point indicated Lee's Corps was found to be engaged, and in need of assistance. Accordingly Walthall's Division was moved out (Loring's following as support), and formed on Lee's left. It attacked the enemy, strongly posted on a hill, and failing after a desperate fight and heavy loss to dislodge him, Loring's Division was placed in position along the Lick-Skillet road, and Walthall directed to withdraw his in rear of Loring's. A short time previous to this General Loring was wounded, leaving his division under the command of Brigadier General Featherston. While his division was taking its position I was myself disabled, and did not return to duty till the 15th of August. At this time the corps occupied the line from the Marietta road to a short distance west of the Lick-Skillet road. Nothing of special interest occurred beyond the constant skirmishing along the picket lines, until the end of August, when the enemy moved from our front, and moved upon Jonesboro'. This corps remained in position around Atlanta until it became necessary to evacuate the place, retiring towards McDonough, and finally to Lovejoy's Station, where it remained until September 18th. On that day we marched for Palmetto, on the Atlanta and West Point Railroad, and on the 20th took position on the left of the Army, between the railroad and the Chattahoochee, where we remained undisturbed until the 29th, when we crossed the river at Pumpkin Town, near Cross Anchor.

It is due them to express my high appreciation of the conduct and services of the several members of my staff, namely, Lieutenant Colonel F. F. Sevier, assistant inspector general, and his assistants, Lieutenants Cohal and Hopkins, and private Williams, of the New Orleans Light Horse; Major Douglas West and Captain W. D. Gale, of adjutant general's department; Major Foster, Captain Porter, Lieutenant De Saullet and McFall, of the engineers; Lieutenants Ridley and Stewart, aids; Captain Vanderford, ordnance officer; Major Mason, quarter master, and Major Murphy, chief of staff. To Captain Greenleaf and his company, the Orleans Light Horse, I acknowledge my obligations for valuable services.

Very respectfully, Colonel, your obedient servant,

(Signed,) ALEXANDER P. STEWART, *Lieutenant General.*

Reports from Loring's Division and from Major General French of action, July 20th, forwarded with this. Others will be forwarded when received. Respectfully,

(Signed,) A. P. S.

OFFICIAL REPORT OF MAJOR GENERAL G. W. SMITH, COMMANDING GEORGIA MILITIA, AT AND NEAR ATLANTA.

HEADQUARTERS GEORGIA MILITIA,
MACON, GEORGIA, *September 15th, 1864.*

GENERAL J. B. HOOD, *Commanding Army of Tennessee, near Lovejoy Station.*

GENERAL:—I have the honor to make the following report of the part taken by the Georgia militia, under my command, during the operations at or near Atlanta. My appointment was dated the 1st of June. I took command a few days thereafter, relieving General Wagner, who returned to the duties of his office as adjutant and inspector general of the State. The force then in the field was composed entirely of State officers—civil and military. They had been formed into two brigades, of three regiments each, and one battalion of artillery, making in all a little over three thousand (3000) men. The officers of the militia not needed for these regiments took their places in the ranks as privates with the civil officers. The command had reported to General J. E. Johnston for duty, and had been ordered to guard the crossings of the Chattahoochee river from Roswell Bridge to West Point, which duty they continued to perform until ordered by General Johnston to cross the Chattahoochee and support the cavalry on the left wing of his Army, the right wing being at Kennesaw Mountain. In the execution of these orders the militia were twice brought in conflict with largely superior forces of the enemy's infantry. They behaved well, thoroughly executed the part assigned to them, and when the Army fell back to the Chattahoochee they were the last infantry withdrawn to the fortified position. General Johnston, in a letter to Governor Brown, paid a handsome, and, I think, a well deserved compliment to them for their conduct beyond the river, and their services in beating back the enemy in their attempts upon the various crossings. The day we marched to the Chattahoochee, we were assigned to your corps of the Army. You soon placed us in reserve, which, it was thought, would give some opportunity for drilling and disciplining the command, no opportunity having offered for this previously. In the meantime the reserve militia of Georgia was ordered out by Governor Brown, and I was ordered to Poplar Spring, near the Atlanta and West Point Railroad, for the purpose of reorganizing, arming the reserves, etc., etc. We had not been there three days before you found it necessary to order us into the trenches on the east side of

Atlanta. You had in the meanwhile been assigned to the command of the Army, and instructed me to report to you direct, instead of through a corps commander. There were at this time about two thousand (2000) effective muskets in the command. We guarded some two miles of the line, having in them, however, about eight pieces of Confederate artillery. On the 22d of July, while Hardee was attacking the enemy on our extreme right in the direction of Decatur, when you ordered the troops on my left to advance, without waiting for orders, I closed the intervals in my line, formed line of battle in the trenches, and moved the militia forward over the parapets more than a mile against the enemy's strong works in our front. They were directed upon a battery which had annoyed us very much. Captain Anderson, who had served with my command beyond the Chattahoochee, volunteered to move his battery with us. He took position in clear open ground within about four hundred yards of the embrasure battery of the enemy, supported by the militia upon his right and left. Within two minutes the effective fire of the enemy was silenced in our front, and after this they only occasionally ventured to show themselves at the embrasures, or put their heads above the parapets. My troops were eager to be allowed to charge the battery, but the brigade upon my left had given way, and the falling back was extending still further to the left. Hardee's fire on my right had ceased just after we moved out of the trenches. I considered it useless to make an isolated attack, and, therefore, held the position and awaited further development. In about two hours I received orders from you, directing me to withdraw to the trenches. We lost only about fifty men killed and wounded. The officers and men behaved admirably. Every movement was promptly and accurately made; there was not a single straggler. A few days after this affair of the 22d of July I was ordered again to Poplar Spring, but was scarcely established in camp before we had again to be placed in the trenches on the left of the Marietta road, and from this time until the end of the siege continued under close fire, night and day. We had to move from one portion of the line to another, and had our full share of all the hardest places, extending from the left of the Marietta road across the Peach Tree road to our extreme right.

The militia, although poorly armed, very few having proper equipments, more than two-thirds of them without cartridge boxes, almost without ambulances or other transportation, most of the reserves never having been drilled at all, and the others but a few days, all performed well every service required of them during an arduous and dangerous campaign. They have been in service about one hundred days, during at least fifty of which they have been under close fire of the enemy mostly night and day. They have always shown a willing spirit, whether in camp, on the march, working on the fortifications, guarding trenches, or upon the open battle field. They have done good and substantial service in the cause of their country, and have established the fact that Georgia is willing and able to do something effective in her own name, besides furnishing more than her quota to the Confederate Armies proper. The greatest number of effective muskets in the trenches at any one time was about five thousand. When Atlanta was evacuated the reserve artillery of the Army passed out through my lines, and my men were formed as a rear guard. The whole was safely brought to Griffin under your orders. The march from Atlanta to Griffin satisfied me that men over fifty are not, as a class, fitted for military duty. I have, therefore, strongly advised the Governor to withdraw them from continuous service.

There being a lull in active operations, the Governor has, with my recommendation and your concurrence, withdrawn the Georgia militia from Confederate service, and furloughed them for thirty days. This report is hastily written, without access to the records and papers of my adjutant general's office, but all omissions can be readily supplied by the returns, etc., already forwarded to your office. Before closing, I cannot refrain from alluding to a subject which, under ordinary circumstances, forms no part of the reports of subordinates to their commanders. I allude to the outcry of the press and people against yourself because of the evacuation of Atlanta.

Unsolicited by me, without my consent or knowledge, the civil and military officers of the State of Georgia, when called upon to take up arms in defence of their homes, almost unanimously elected me their leader, and as their leader I wish, in this report, to say to you and place officially on record this opinion, viz: Had your orders been properly executed, either upon the 20th of July, at Peach Tree creek, the 22d of July, on our right, or on the 30th of August, at Jonesboro', Sherman could have been foiled and Atlanta saved at least for some time to come, and I am not alone in this opinion.

Commanding a peculiar organization, the ranking officer of the forces of the State in which you were operating, I was invited to and participated in your councils. I had every opportunity of knowing what was going on. Your plans were fully explained to your lieutenant generals and your chief of artillery, chief engineer, and myself. Opinions and views were called for, and then specific orders were given. I have never known one of them to dissent to any plan of yours that was attempted to be executed, never a doubt expressed as to the meaning and intent of your orders, nor a suggestion made by them of a plan they supposed would be better than that you ordered. If they are not now unanimous, there is but ONE, if any, who dissents from the opinion expressed above, viz: Sherman would have been beaten had your orders been obeyed on the 20th of July, 22d of July, and 30th of August.

Whatever the press or people may say, the militia of Georgia are now more than satisfied with you as their Confederate General, and when they again enter that service in defence of their homes will be glad to hail you as their Confederate Chief.

(Signed) G. W. SMITH,
Major General.

LETTER TO HONORABLE MR. SEDDON.

HEADQUARTERS, ARMY OF TENNESSEE,
Near Nashville, Dec. 11th, 1864.

HON. JAS. A. SEDDON,
Secretary of War, Richmond, Va.

SIR:—On the 21st of November, after a delay of three weeks, caused by the bad condition of the railroad from Okolona to Cherokee, and of the dirt road from the latter point to Florence, and also by the absence of Major General Forrest's command, this Army moved forward from Florence—Major General Cheatham's Corps taking the road leading towards Waynesboro', and the other two corps moving on roads somewhat parallel to this, but more to the eastward, with the cavalry under General Forrest in their advance, and upon their right flank.

The enemy's forces at this time were concentrated at Pulaski, with some force also at Lawrenceburg. I hoped to be able to place our Army between these forces of the enemy and Nashville, but he evacuated Pulaski upon the 23d, hearing of our advance (our cavalry having previously driven off their forces at Lawrenceburg), and moved rapidly by the turnpike and railroad to Columbia.

The want of a good map of the country, and the deep mud through which the Army marched, prevented our coming up with the enemy before they reached Columbia, but on the evening of the 21th of November our Army was placed in position in front of the enemy's works at Columbia. During the night, however, they evacuated the town, taking position on the opposite side of the river, about a mile and a-half from the town, which was considered quite strong in front. Therefore, late in the evening of the 28th of November, General Forrest, with most of his command, crossed Duck river a few miles above Columbia, and I followed early in the morning of the 29th with Stewart's and Cheatham's

Corps, and Johnson's Division of Lee's Corps, leaving the other Divisions of Lee's Corps in the enemy's front at Columbia. The troops moved in light marching order, with only one battery to the corps. My object being to make a rapid march on roads parallel to the Columbia and Franklin pikes, and by placing the troops across the pike at or near Spring Hill, to cut off that portion of the enemy between Spring Hill and Columbia. The cavalry engaged the enemy near Spring Hill about midday, but their trains were so strongly guarded that they were unable to break through them. About 4 p. m. our infantry forces, Major General Cheatham in the advance, commenced to come in contact with the enemy about two miles from Spring Hill, through which the Columbia and Franklin pike passes. The enemy were, at this time, moving along the pike, with some of their troops formed on the flank of their column to protect it; Major General Cheatham was ordered at once to attack the enemy vigorously, and get possession of this pike; and although these orders were frequently and earnestly repeated, he made but a feeble and partial attack, failing to reach the point indicated.

Darkness soon came on, and to our mortification the enemy continued moving along this road almost in ear-shot, in hurry and confusion, nearly all the night, Thus was lost the opportunity for striking the enemy for which we had labored so long, the best which this campaign has offered, and one of the best offered us during the war. General Cheatham has frankly confessed the great error of which he was guilty, and attaches much blame to himself. While his error lost so much to the country, it has been a severe lesson to him by which he will profit in the future. In consideration of this, and of his previous conduct, I think that it is best that he should retain for the present the command he now holds.

Before daylight, next morning (30th November), the entire column of the enemy had passed us, retreating rapidly towards Franklin, burning many of their wagons. We followed as fast as possible, moving by the Columbia and Franklin pike, Lieutenant General Lee, with his two divisions, and trains and artillery moving from Columbia by the same road. The enemy made a feint of making a stand in the hills, about four miles from Franklin, in the direction of Spring Hill, but as soon as our forces commenced deploying to attack them, and extending to outflank them on their left, they retired slowly to Franklin. This created a delay of some hours. We, however, commenced advancing on Franklin, and attacked the place about 4 p. m., with the corps of Generals Stewart and Cheatham—Johnson's Division of Lee's Corps becoming engaged later. We carried the enemy's outer line of temporary works, but failed to carry the interior line. During the night I had our artillery brought forward, and placed in position to open upon them in the morning when the attack should be renewed, but the enemy retreated rapidly during the night on Nashville, leaving their dead and wounded in our hands. We captured one thousand prisoners, and several stands of colors. Our loss in officers was severe. The names of the general officers I have already given by telegraph;* our entire loss was about forty-five hundred.

We continued our march towards Nashville, and on the 2d of December our Army took its present position in front and about two miles from the city.

Lieutenant General Lee's Corps, which constitutes our centre, rests upon the Franklin pike, with General Cheatham upon his right and General Stewart upon his left. Our line is strongly entrenched, and all the available positions upon our flanks and in rear of them are now being fortified with strong self-supporting detached works, so that they may easily be defended should the enemy move out upon us.

The enemy still have some six thousand troops strongly entrenched at Murfreesboro'; this force is entirely isolated, and I now have the larger part of the cavalry under General Forrest, with two brigades of infantry, in observation of these forces, and to prevent their foraging on the country. Should this force attempt to leave Murfreesboro', or should the enemy attempt to reinforce it, I hope to be able to defeat them.

I think the position of this Army is now such as to force the enemy to take the initiative. Middle Tennessee, although much injured by the enemy, will furnish an abundance of commissary stores, but ordnance and certain quarter master stores will have to come from the rear, and, therefore, it is very important that the railroad should be repaired at once from Cherokee to Decatur; the cars can run now from here to Pulaski on

---

*HEADQUARTERS, ARMY OF TENNESSEE,
SIX MILES SOUTH OF NASHVILLE,
December 3d, 1864.

HON. J. A. SEDDON,
Secretary of War, Richmond, Va.

SIR:—About 4 o'clock, p. m. we attacked the enemy at Franklin, and drove him from his outer line of temporary works into his interior lines, which he abandoned during the night, leaving his dead and wounded in our possession, and retreated rapidly to Nashville, closely pursued by our cavalry. We captured several stands of colors, and about one thousand prisoners. Our troops fought with great gallantry. We have to lament the loss of many gallant officers and brave men. Major General Cleburne, Brigadier Generals Gist, John Adams, Strahl, and Granberry were killed; Major General Brown, Brigadier Generals Carter, Manigault, Quarles, Cockrell, and Scott were wounded, and Brigadier General Gordon captured.

(Signed) J. B. HOOD, *General.*

the Tennessee and Alabama Railroad. We have sufficient rolling stock captured from the enemy to answer our purposes. I will endeavor to put this road in order from Pulaski to Decatur as soon as possible.

As yet I have not had time to adopt any general system of conscription, but hope soon to do so, and to bring into the Army all men liable to military duty.

Some fifteen thousand of the enemy's Trans-Mississippi troops are reported to be moving to reinforce the enemy here. I hope this will enable us to obtain some of our troops from that side in time for the Spring campaign, if not sooner.

Very respectfully, your obedient servant,

J. B. HOOD, *General.*

Made in United States
North Haven, CT
04 November 2026

21766790R00202